GOVERNING BODIES

# GOVERNING BODIES

A MEMOIR, A CONFLUENCE, A WATERSHED

# SANGAMITHRA IYER

MILKWEED EDITIONS

© 2025, Text by Sangamithra Iyer
All rights reserved. Except for brief quotations in critical articles or reviews, no part of this book may be reproduced in any manner without prior written permission from the publisher: Milkweed Editions, 1011 Washington Avenue South, Suite 300, Minneapolis, Minnesota 55415.
(800) 520-6455
milkweed.org

Published 2025 by Milkweed Editions
Printed in Canada
Cover design by Mary Austin Speaker
Cover illustration by F. O. Finch
Interior art by Lavanya Manickam;
photograph on page 354 by Lisa Hirmer
Author photo by Jonathan Blanc
25 26 27 28 29   5 4 3 2 1
*First Edition*

978-1-57131-393-5

Library of Congress Cataloging-in-Publication Data has been applied for.
LCCN: 2025002022.

Milkweed Editions is committed to ecological stewardship. We strive to align our book production practices with this principle, and to reduce the impact of our operations in the environment. We are a member of the Green Press Initiative, a nonprofit coalition of publishers, manufacturers, and authors working to protect the world's endangered forests and conserve natural resources. *Governing Bodies* was printed on acid-free 100% postconsumer-waste paper by Friesens Corporation.

*Begin again.*

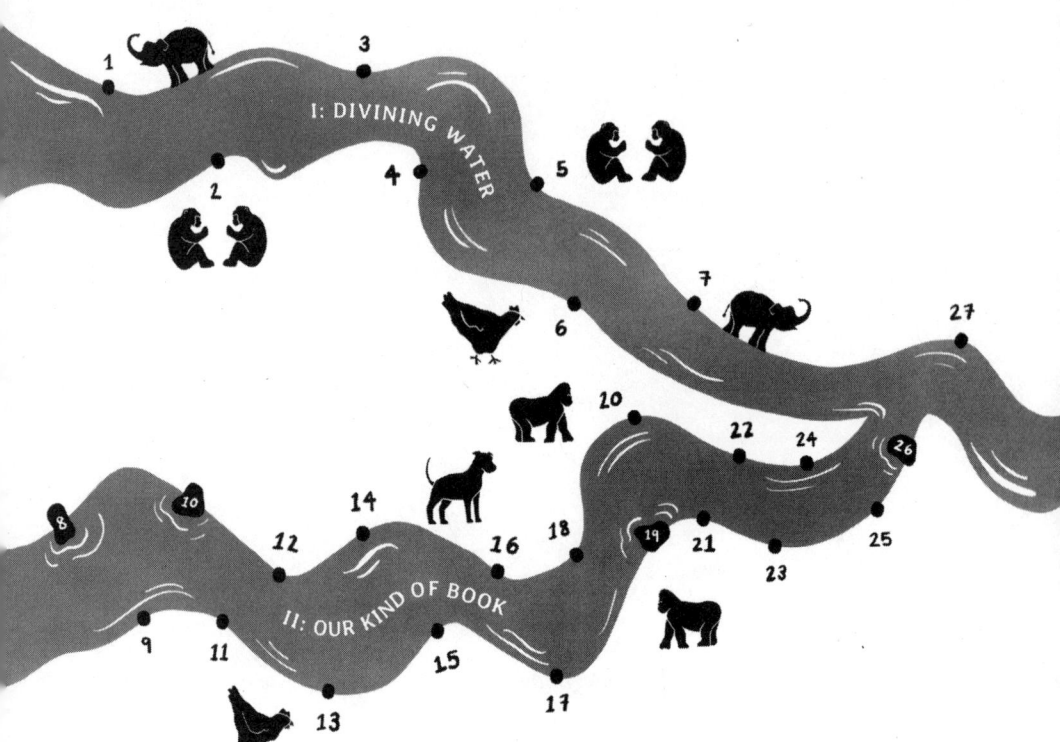

## CONTENTS

### ❈ I: DIVINING WATER

1. Divining Water ... 5
2. On Bridges ... 17
3. Soil Mechanics ... 25
4. In Situ ... 31
5. Small Small Redemption ... 35
6. Pile Driving ... 47
7. Burmese Years ... 55

### ❈ II: OUR KIND OF BOOK

8. Question Mark ... 69
9. Fingerprint of Water ... 71
10. Air ... 81
11. Without Sorrow ... 85
12. Truth ... 93
13. The Chicken Issue ... 103
14. A Thing After Your Own Heart ... 109
15. Love Is Work Made Visible ... 127
16. Eyes Are Sympathetic ... 131
17. An Attitude of Gratitude ... 133
18. Work ... 137
19. THINK-TROUBLE ... 141
20. Soiled Hands ... 143
21. Bearing Capacity ... 155
22. Finding Kallakurichi ... 157
23. Returns ... 175
24. Catena ... 179
25. Verge ... 181
26. The Truth Is in the Body ... 207

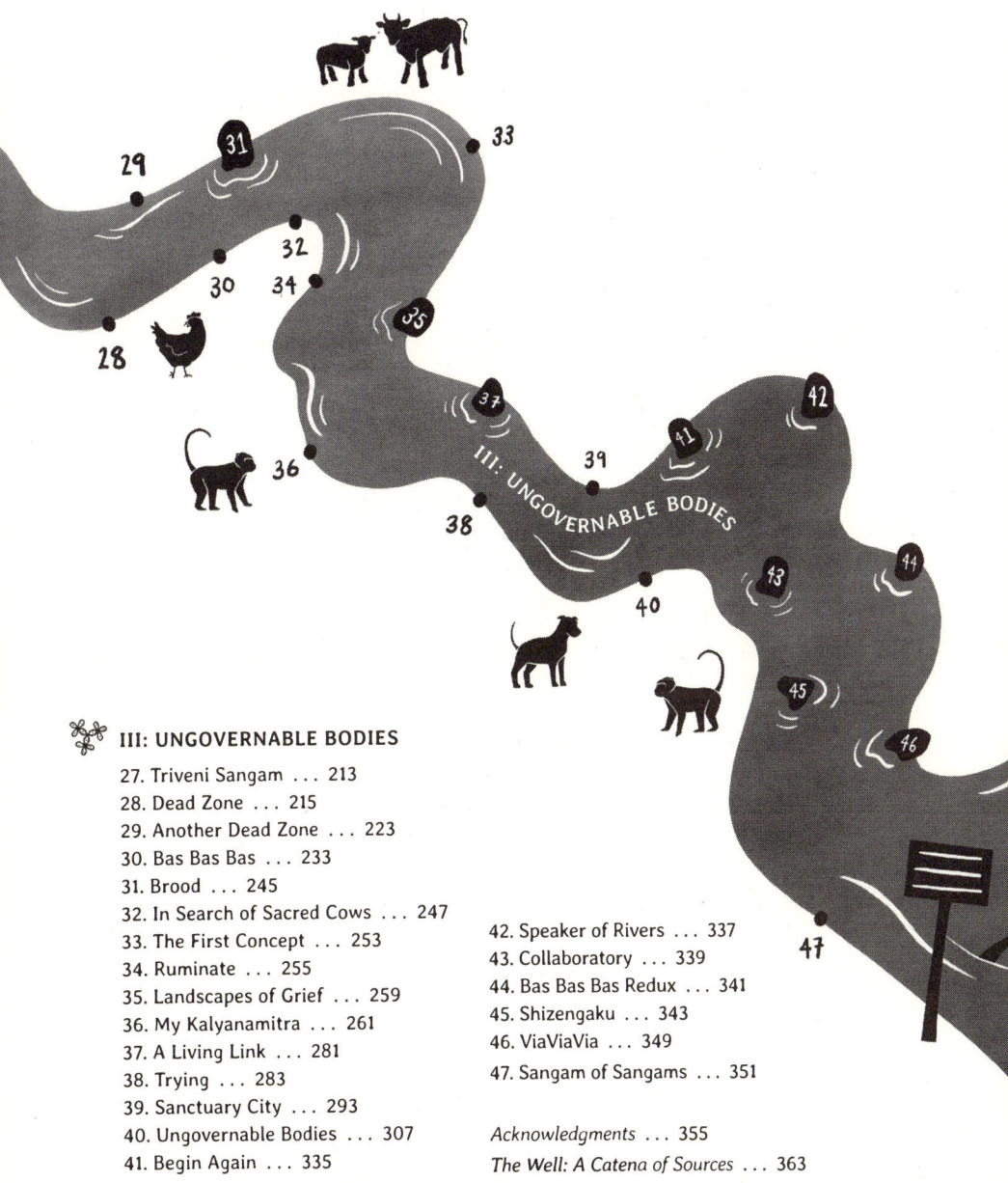

✺ **III: UNGOVERNABLE BODIES**
  27. Triveni Sangam ... 213
  28. Dead Zone ... 215
  29. Another Dead Zone ... 223
  30. Bas Bas Bas ... 233
  31. Brood ... 245
  32. In Search of Sacred Cows ... 247
  33. The First Concept ... 253
  34. Ruminate ... 255
  35. Landscapes of Grief ... 259
  36. My Kalyanamitra ... 261
  37. A Living Link ... 281
  38. Trying ... 283
  39. Sanctuary City ... 293
  40. Ungovernable Bodies ... 307
  41. Begin Again ... 335
  42. Speaker of Rivers ... 337
  43. Collaboratory ... 339
  44. Bas Bas Bas Redux ... 341
  45. Shizengaku ... 343
  46. ViaViaVia ... 349
  47. Sangam of Sangams ... 351

*Acknowledgments ... 355*
*The Well: A Catena of Sources ... 363*

And to trust that all water touches all water everywhere.
—ALEXIS PAULINE GUMBS

# GOVERNING BODIES

What you must understand

    is that when I tell you a story about my body,

        I cannot separate it from a story about water.

    And a story about water is also a story about family.

    And a story about family is rooted in the earth.

And when I tell you about the earth,

    I must tell you about elephants and chimpanzees,

        cows and chickens,

        coral and trees,

        monkeys and bees.

    What harms one body harms all bodies.

Like tributaries to the same river, our stories are entwined.

# I

# DIVINING WATER

*For Thatha*

All water has a perfect memory and is forever trying to get back to where it was.
—TONI MORRISON

# DIVINING WATER

T̲HE IRRAWADDY RIVER IN BURMA IS NAMED AFTER THE mythical, multitrunked, white elephant Airavata, derived from the Sanskrit word Iravat, *one who is produced from water.*

Our family history is also a story produced from water. If I were to trace your engineering career, Thatha, I'd follow it down the Irrawaddy River. If I were to trace mine, I'd follow it from streams in the Catskill Mountains through aqueducts and tunnels to New York City's pipes and faucets. My experience is also in the Yosemite Valley–Sierra Nevada snowmelt that gravity carries to San Francisco. It is on rooftops and in rain barrels in Cameroon; in buckets in the Sanaga River.

Our lifetimes do not overlap, yet these lines of Alfred, Lord Tennyson's "The Brook" are passed down in our family from you, my grandfather, to my father Adi, your youngest child, to me, his youngest daughter.

> And out again, I curve and flow
> to join the brimming river,
> for men may come and men may go,
> but I go on forever.

The story I heard growing up was that you worked for the British in Burma before quitting to become a satyagrahi and live a life of self-reliance in the Tamil Nadu village of Kallakurichi.

Kallakurichi is an origin story, a beginning, a chosen ancestral land, an idea, a moral philosophy, a refuge, a refusal. Kallakurichi is a returning memory—my father lost in thought, with a private smile, an inside joke. Kallakurichi is a dream. Yours, maybe mine. Kallakurichi lives in the body. Yours, maybe mine.

I see it as your attempt to practice nonviolence while the society around you is teeming with oppression, both from colonialism and the caste system. In the 1930s, you trade in your civil engineering post to become a water diviner, part of a call to action to develop wells for communities who were denied access to water because cruelty deems them untouchable. You understand that water and sanitation are closely linked to social justice. I understand this too.

My father—the youngest of your thirteen children—is born into what I think is this utopian experiment in Kallakurichi. Is *utopia* the word you would have used?

Revolution begins by learning to spin. Your children all wear homespun khadi clothes like you do. Education is learning the landscape, climbing trees, drawing water, and learning poetry.

My father grows up studying Sanskrit and English in addition to speaking Tamil at home. Each day, you order him to copy a page of the dictionary by hand and memorize Shakespeare, Tennyson, Goldsmith, and others "by heart" from the poetry collection you share with the children, which ultimately makes its way to the youngest sibling. This worn copy of *Memory Work and Appreciation* is also gifted to me, many years later and oceans away. I, too, learn these same lines by heart. This book of British verse is the only physical artifact passed on between our three generations. Why is there not anything else? Your letters? Your divining rod?

Initially, I find it curious how you, who quit the British in Burma to join the Freedom Movement in India, showered your children with the oppressor's literature. But I now realize that relationships are complicated, and perhaps noncooperation need not apply to poetry.

When I am twenty-five years old, I again recite those lines of Tennyson by heart to my father in New York, at Good Samaritan Hospital—the same hospital where I was born. I don't know if he hears me or understands that he is dying.

My father leaves the world from the same place I enter it.

*Apostrophe* comes from the Greek for *to turn away*. Literary apostrophes are a way of writing to an absent presence. Apostrophes are a mark of what is missing, an indication of possession, or speech within speech. I am turning toward absence, toward you Thatha, toward what is both missing and possessed.

"The Irrawaddy River has five defiles and forty-six constraints," another engineer and water activist in Burma tells me, eighty years after you have left. "Defile means that the river is narrow because of the bedrock," she explains. "Constraint means you have an island in the middle, or the riverbank."

I think about the defiles and constraints in my storytelling. There are so many things I want to say that cannot fit through the narrow corridor of a defile. I will also make so many stops along the way. Maybe these constraints are more like guideposts, places to rest and reflect.

I have been following water, and like a log floating down the Irrawaddy, it has taken years for me to make my way to the mouth. What have I come to say?

# GOVERNING BODIES

In the weeks following my father's death, I am searching more and more for meaning, purpose, anything. It comes in bursts in unexpected places. There is the pounding of the pile driving hammer on a construction project I am monitoring in Flushing, Queens. With each hammer blow, the earth shakes, and my heart skips a beat. I suddenly become aware of my body, hidden behind baggy clothes and a clipboard. My vision and hearing are dampened by goggles and earplugs, but I can still hear the chickens at the neighboring live market—bodies expressing themselves until their very last moments. I know that in the split-second drop of the pile driving hammer, their lives can end. I am not able to do anything for those birds. I walk by quickly and look away.

Around this time, I write about my recent experience volunteering with orphaned chimpanzees in Cameroon, for *Satya*, a magazine dedicated to animal advocacy, environmentalism, social justice, and vegetarianism. The founders of this New York City–based magazine chose the Sanskrit word for *truth* as its name because they were inspired by satyagraha, the same movement you left your engineering career to join.

I will tell you more about the small smalls, the rescued baby chimpanzee girls in my care. When their tiny fingers and toes cling to my body, I can feel their loss and longing. At this moment, I am like you—figuring out how to leave engineering for activism.

I write that story for my father because I share his respect for animals and love of words. It is the first story I publish; one he will never read.

There is a sepia-toned photograph of your son at Jones Beach from 1971. He stands on his head with his legs bent in the lotus position. East meets West, and the world is upside down. As a new arrival in the United States, my father befriends and teaches yoga to hippies in New York City. His body in this picture is slender and nimble, unlike his body in the hospital in 2003, the one that is later burned to ash and released into the coliform-infested waters of the Ganga.

Your youngest daughter, my Jaya Aunty, sees my father in my face. "Carbon copy," she says. "Adi moonji." Adi's face. "Adi ditto." It's more than our cheeks and chin clefts that are similar. It's the ahimsa. "Adi romba sensitive," Jaya Aunty says. *Very* sensitive. He is only sixteen years old when you leave the earthly plane. This ahimsa, this absence of a desire to harm, is often translated as nonviolence, but I haven't found an equivalent word in the English language that captures its full meaning. Have you?

Jaya Aunty remembers women passing by Kallakurichi carrying heavy burdens of firewood on their heads. My father, though small and skinny, would offer to share some of the load and walk with them on their journey. He despised the firecrackers the other children enjoyed at Diwali time. He knew how the sounds upset the monkeys in the trees.

I remember my father always taking care of injured wildlife near our home in the suburbs of New York City. When we were young, my parents took my brother and me to see this movie called *Return of the Jedi*. My father was so disturbed by the sight of the character named Jabba the Hutt eating a live frog, he left the theater. His death certificate says *cardiac arrest*, but I know my father's heart suffered from extraordinary empathy.

My body does not remember the yoga my father taught me, the same yoga you probably taught him. After his death, I enter a Vinyasa yoga class at a YMCA in New York City. *Flow* is what my acupuncturist tells me my grief-ridden body needs—to restore my flow. The yoga instructor, a young white male, clasps his hands and bows at me. "Namaste," he says. He tells us that if we are unable to keep up, to "just go back to child's pose." I am too embarrassed to ask which one is child's pose. My body eventually remembers. I collapse my torso over my bent knees.

At the time of the Jones Beach photograph, my father has been in the United States for two years. He arrives with only seventy-five cents in his pocket. In India, he studied social work, wrote poetry, and dreamed of being a journalist or working in foreign service. You homeschool them, and send them to college young. My father is only thirteen years old when you enroll him in college. Your eldest sons pursue engineering, like you. And your youngest son, born in Kallakurichi, pursues social work, like you too. That is what you call yourself in Kallakurichi, a Harijan Welfare Worker, a social worker.

"Adi was born a poet," Jaya Aunty tells me.

I collect these descriptions of my father and of you that people mention in passing. Each is a little gift, to be able to get closer to you both after you are gone.

My mom shared with me stories I had never heard when my father was alive. After he finishes his master's degree in social work, she tells me, my father lives with your eldest son, Sundaresan Uncle in Delhi, and gets a sales job during the day and moonlights tutoring English to the wife of the Congolese ambassador to India. A car picks my father up, and the ambassador's chef cooks my father chapatis while he gives English lessons. Who are these people who knew my father before I did?

When he first arrives in the United States, my father lands a job as a psychiatric social worker at Johns Hopkins Hospital in Baltimore, where he teaches yoga to one of his patients. Later, that student will give up meat, raise two vegetarian dogs, and write a booklet, *To Become a Yogi*, and dedicate it to my father. I find it one day in high school in a filing cabinet my parents keep in their puja room. There is one passage that is etched into my memory: "When we eat an animal that has been killed for our benefit, we awaken the killing instinct in ourselves and gradually lose respect for all life." It gives me language I do not yet have, for knowledge I do.

In New York in the 1970s, my father becomes a social worker at Manhattan Psychiatric Center on Wards Island, a small landmass in the waters between Manhattan and Queens, that the Indigenous Lenape call Tenkenas, or *wild lands*.

Over thirty-five years after my father leaves his job on Wards Island, a colleague tells me about the time she was stationed at the sewage treatment plant on Wards Island in the 1980s. "After a while, you get used to the smell," she says. "You don't even notice it anymore." Back then, there were wild dogs all over the island. "You have to run to your car at night to avoid them," she tells me. "Then one year, they were all gone."

What was the thinking behind placing a mental health hospital and a wastewater treatment plant in close vicinity to one another? Does my father think about the connections between sanitation and social justice on Wards Island like you do about access to water in Kallakurichi?

The Clean Water Act wasn't enacted until 1972. Does it bother him, the proximity to the sewage treatment plant? I wonder what my father smelled back then. Does he know that the island is also the site of a former cotton mill? I never formulate such questions when I have the opportunity to ask him. I now imagine my father on his lunch break, feeding the wild dogs in these wild lands and looking out onto the East River.

"I thought you followed the elephant," my mother tells me. She is recounting a story from when we move back to India for two years after I was born while she pursued her master's degree, also in social work. My first language is Tamil, one I largely lose when we return to the United States. One day during this time, I disappear. It is January, and an elephant adorned with garlands of marigolds and graffitied with sandalwood paste marches through the streets as part of a temple procession. The elephant's human handlers guide her to each house. Families offer bananas to the auspicious pachyderm believed to be a stand-in for a god. She is trained to place her moist snout on the devotees' heads—a tickle perceived as a blessing. She leaves behind a trail of droppings the size of coconuts.

After the elephant reaches my maternal grandparents' home in Bangalore—where we are living at the time—I vanish. My mother assumes that since I, like my father, have a fondness for animals, I must have followed the elephant. She and her brother, Ramani Mama, run around the neighborhood trailing the procession, asking if anyone has seen a small girl wearing a striped T-shirt and a pair of jeans with short hair closely cropped to her head. My head was recently shaved, shedding my past life.

They eventually find me at a neighbor's house. After observing the largest creature I have ever seen in my life, I go over to tell them that there is an elephant in the street decorated just like the elephant statue they have displayed on their shelf in their home. Does the miniature suddenly seem insufficient—unable to move ears, trunk, and tail in the same manner as the live one?

I don't remember this event. But I try to imagine my toddler self, summoning another past that was shed. My little arms extend wide approximating the size of the elephant, and then I point to the small replica. A discovery. A story I tell in Tamil, a tongue I no longer command except for a few words like Thatha, *grandfather*, Patti, *grandmother*, thanni, *water*, and yaanai, *elephant*.

I am raised worshipping the elephant-headed god, Ganesha. I am taught to cross my arms over my chest, pull on my ear lobes, and squat-bow to the half-pachyderm deity, the remover of obstacles. I learn the story of how Ganesha gets his face. Lord Shiva came home one day to a child whom his wife, the goddess Parvati, made from clay. Parvati was having her bath when Shiva arrived, and her child obeyed his mother's commands, and would not let this stranger, his father, into the home. Shiva, not knowing who this clay child was, beheaded him in a fit of anger. When he realized what he had done and the grief it brought to Parvati, Shiva ran into the forest and took the head of the first creature he encountered—a baby elephant—to bring the boy back to life. I don't understand how a god can behead two beings. The boy was resuscitated as Ganesha, but I ask, "What happened to the baby elephant?" This, Thatha, is the first of many questions I have about faith. Still, I am curious about this elephant-headed boy—comprised of two beings deprived of their whole.

In the outskirts of New York City, where I grow up, my parents recreate the foods from their homeland. I delight in soft fluffy idlis with sambar with radishes. I skim the clear liquid of peppery rasam and drink it like a soup. My brother Manu and I love kunthaloo, a vegetable with no English name that I am aware of, cooked slightly burnt and crunchy. The aroma of cumin and coriander permeates our home, seeps into my clothes and backpack, and is transported into my locker at school.

"What's that smell?" a schoolmate asks one day, as we grab our books between classes.

"I don't know," I say, embarrassed, as I shut my locker door and rush off to class.

One day in elementary school, I forget my packed lunch, and my teacher decides to buy me food in the cafeteria. They are serving hamburgers that day. While waiting in the cafeteria line, I motion to my teacher to bend down so I can tell her something. I whisper to her that I don't eat meat.

"What do you want on your bread, then?" she yelps.

"Lettuce." Another whisper.

"What else besides lettuce?" she asks.

"Nothing," I say.

"You can't eat just that. That's rabbit food!" she says.

The little kids in line erupt with laughter. Despite my desire to be perceived as "American" at this age, I never wish to eat meat, but I don't have the words to describe why not when bombarded with questions.

"Don't you want to know what it tastes like?" kids ask me.

"You never ate meat in your life?"

"Not even by accident?"

Back then, there are many things about myself I am discovering, many things I am uncertain about. As the child of immigrants, I am reconciling my identity between two worlds, neither to which I feel I fully belong. But not eating animals is one thing I am sure of even if it alienates me further from my peers.

It is sometime in middle school when I make it clear that this is my choice, not a family or religious obligation. A sense of justice with a shift in diction. A change in two letters.

"You can't eat meat?"

"I *won't* eat meat."

I think back to this time, and I see a child having to witness a violence commonly accepted but rarely questioned. She refuses to participate.

A few years after my father's death, I accompany your second-born son, Amirthalingam Uncle, to a temple in Tamil Nadu. Adjacent to the temple hall is a medium-sized room with a separate entrance. The gates to this area open, and I am surprised to see an elephant—just barely smaller than the room she is enclosed in—walk out. Devotees pay money to feed the elephant—a business scheme for the religious enterprise. No one suspects a Hindu temple to be guilty of animal cruelty.

A man with a stick stands next to the elephant to control her. I attempt to use my limited Tamil to interrogate the handler, but I can bark only a few basic questions. *Eat? Sleep where? Children?*

I want to channel that younger me with her near-shaven head, Tamil fluency, and astute observations. I can see her again extend her arms wide and then point to the elephant's enclosure. She can say all that I cannot, all that my father could have said if he was still alive and there with me. How can you restrict this wild animal to this small space? Why must you control her body?

I realize there are so many things I don't know about my father's past, which suddenly feel urgent as I am figuring out how to live in an unjust world. I want to better understand where I come from, which is intimately tied to understanding both my father and you. I am interested in how people defy what is expected of them to pursue something purposeful, to seek larger connections. I want to understand how we pursue satya and practice ahimsa in the modern world.

I know you became a water diviner in India after leaving Burma. I love that word, divining. If I can locate water, I think, I can trace it to you, to my father, and perhaps to me. But what am I really searching for? This desire to dowse is a desire to restore some lost ancestral knowledge, some intergenerational memory.

I studied groundwater hydrology in graduate school and oversaw the drilling of wells. It may sound strange that I would resort to a tree branch to locate water, uncover family history, or divine truths. But you are a civil engineer turned water diviner too. Maybe you are the only one who understands. I picture myself holding the arms of a Y-shaped tree branch and feeling the force of water pulling me in one direction. Where will it take me?

Can the divining rod guide me to all of your engineering posts in Burma, from the Irrawaddy's source at the confluence in Myitsone down to the small lighthouse island where the Tavoy River empties into the Andaman Sea? Can the divining rod let me feel the blood rushing through your body, the passion and conviction of your decision to quit the British, rid yourself of material wealth, and pursue a life of service aligned with your ethics back in Tamil Nadu? Does it lead me to that bend in the Gomukhi River where my father was born? Or to where I leave his ashes swirling in the Ganga?

Even without that ancestral rod, perhaps I have been divining water all these years as an engineer and activist. I follow the flow from source water reservoirs to the outfalls of wastewater treatment plants. I sketch the drainage patterns of rainwater in a forest of chimpanzees. Water is my guide along the plastic-choked Yamuna, carrying the blood of chickens and the excrement of no-longer-sacred cows.

Water bodies, like our animal bodies, are poisoned and controlled. If I follow these rivers, can I divine how our bodies can carry these histories and still be liberated?

This is my story produced from water, my body of work.

# ON BRIDGES

During college, an assignment for Professor Guido sends me looking for your bridge. I don't know what it is called, only that it is in Burma and that you built it. That is what my father remembers. It is my sophomore year at The Cooper Union for the Advancement of Science and Art, and I spend most of my time at 51 Astor Place, the yellow brick building called the Albert Nerken School of Engineering. I don't mean to end up in engineering school. I am good in math and science in high school, but my refusal to dissect and test on animals sheds doubt on my survival in a premed program. My father, who loves literature and history as much as I do, wants me to study something else. "Learn a practical skill," he says. "You can always read books on the side." The logic is this: it's much easier for an engineer to pick up politics than it is for a history major to teach herself physics. As an immigrant who arrived in this country with only a few coins in his pocket, his advice is rooted in survival—he does not want me to experience the debt and poverty he knew. But I want to pursue my passion, although I don't know what that is yet, only that I have it.

I apply on a whim to Cooper Union, but when I get in, I'm not sure if I want to be in New York City, in such a small school, with such a small focus. I think college is where you read books under trees and get arrested for political protests. But I can't justify not going. Cooper Union is free (then). So, I end up in the concrete jungle of the East Village: no trees, no protests (then), but also no tuition.

I want to study something meaningful, so I write *environmental engineering* on my application. Is this my passion?

I have my doubts on engineering. Unlike some of my peers, I don't really want to build anything or destroy anything. I just like to analyze and understand how things work. Certain subjects, such as thermodynamics, don't come intuitively to me, but mathematics is my language. I can manipulate equations and interpret numbers until they reveal a story.

But, is this my purpose? I wonder this many times while sitting on the front steps of the engineering building, looking at the incorrect time displayed on the giant musical-note clock on the Carl Fischer building. I'd hear the skateboarders and their screeching wheels congregating at "The Alamo," the large black cube at the top of Cooper Square.

"Did you know the cube spins?" I tell friends visiting New York.

We would each grab a corner of the black prism and, sure enough, it would turn. There is that one time after the rain, when all the stormwater that accumulated on top of the Alamo falls all over us. "Let's not do this anymore," my friend Mar says.

Sometimes I'd sit and stare at the mice scurrying outside the entrance to the building. My friend Kari likes to think of them as kittens.

"Oh, they are just little kittens," she'd say.

"The rats?" others would inquire.

"Oh, you mean the cats?" she'd reply.

When Professor Guido gives us an assignment to write a report on a structure of our choosing—any structure in the world—I think I can use this opportunity to learn more about your work in Burma. Though we never met, snippets of your life as an engineer and activist make it into my college application essays. When the local paper in Rockland County writes a story on me in high school, it states I am inspired by my grandfather, a civil engineer who was a follower of "Mahatma Gandhi, the Indian moral and civil leader who transformed India and the world with his freedom movement." Back then, I really don't know all that much about you, or Gandhi for that matter.

"I'm going to write about my grandfather's bridge," I tell Professor Guido. Vito Angelo Guido. Professor Guido's specialty is geotechnical engineering, one I later pursue. I am grateful for his introduction to the subject, but when I first met him sophomore year, I was terrified of disappointing him. There was this indelible lecture he gave my class after our poor performance on the first exam. "If you can't get this right, you can't do anything in civil engineering," he said. "You might want to start rethinking your careers." He paused and reconsidered. "I guess you could still do asphalt." I felt sick. Pavement design, I knew, was not my passion.

I came away with two things that semester. First, if I want to, I can teach myself to do anything. And second, there are things, like asphalt, that I will never do.

"What's the name of the bridge?" Guido asks me. He sits at his desk at the front of our classroom, with his notebook in front of him. He looks up at me, waiting for my response.

"I don't know," I tell him. He scribbles something in his log. *Asphalt*, I fear he thinks.

My initial searches for your bridge come up empty. I do not find anything about this structure, not even its name. But this is my first attempt to find you through your work.

A couple of days before my paper is due, I give up. I do my report on Yankee Stadium. I don't even like the Yankees, or baseball for that matter.

After that semester, I abandon this search for your bridge in Burma and take up interest in the New York Liberty. It is 1997 and the Women's National Basketball Association has just started, and I feel so lucky to be in New York City that summer to witness the inaugural season.

I receive a summer fellowship in orthopedic bioengineering at the Hospital for Joint Diseases in Manhattan. My assignment is measuring the effect of acetabular cup orientation on range of motion for artificial hip replacements. At night, I am free to enjoy the city—no classes, no homework. If the Liberty are in town, chances are I will invite whoever is around to come with me to Madison Square Garden. My fellowship stipend is modest, but I can afford Liberty tickets which, back then, are about the cost of going to a movie. Usually my friend, Paul, is down for a game, or my brother, Manu, can join if he is in the City. Basketball is my brother's passion, and growing up he converts our driveway basketball court into a masterpiece work of art called the I-dome. He has a tagline for it too—"Where your dreams become reality and your best shines through." Basketball means more than basketball, to Manu. The I-dome, like Kallakurichi, is embodied with a philosophy. It lives in collective memory and changes over time.

One year, our cousin Anjali, Amirthalingam Uncle's daughter, visits us at the I-dome. She is also an artist and admires Manu's work. It is a basketball court and a museum and a memorial all at once. It is a tribute to his friends and family, and to my father. Manu tells Anjali that it now also feels like basketball was an inheritance from my father and Sayana Uncle, who loved the sport. Anjali tells him it's better to make that connection now, as an adult, instead of having that pressure as a kid.

"You have your genes, you have genetics, you have baggage. All of us have," Anjali says. "But this is *your* thing. There is this Hindi word," she tells us. "It's called junoon. It is *passion*. But it's also *madness*."

We talk about this *junoon* as what Manu has for the I-dome, and his artistic messages that have appeared over time which embody his beliefs. "I always wanted to bring the world together," Manu says of the governing desire of the I-dome. "I think, for me, it gave me great peace to have this place."

Anjali says that she, too, lives her life with junoon. "It's like a passionate madness and that's what makes you do what you want to do."

I think of your junoon in creating Kallakurichi, and what you think of us, your grandchildren, finding ours.

In 1997, during that first season of the WNBA, there are many empty seats, so we can buy the cheap tickets for eight dollars and move down for a better view. I'd carry a couple of teaspoons to cheer my favorite player, Teresa Weatherspoon. At the end of the game, we'd find ourselves near the friends and family section, next to the locker rooms. Loved ones wait for the players to come out. If we are lucky, we meet a player. I carry a picture of the I-dome and an invitation to come visit. One day, I meet my idol T-Spoon and give her that invitation. She says, "You know what, I'll give you a call and we are gonna play some ball." I still wait for that call.

Security was so lax back then. Usually, the guards went home early, and after the players headed out, my companions and I would be the only ones left, and we could walk onto the shiny parquet Garden floor.

Serendipity and possibility are what I remember when I think of that time. Maybe I could become a WNBA historian, I think. I can start from the beginning. Instead of the seemingly impossible task of trying to salvage lost stories from the past, I can record history by documenting the present to the future. This idea doesn't last, and I always come back to the past, to you, to Kallakurichi.

The following semester, in my junior year of college, I discover a different kind of bridge that would change my present and future. It is my junoon. I receive a care package from my parents with *Next of Kin*, a book I mention wanting to read after hearing about it in my Urban Archeology class. They write an inscription: "Dear Sangu, Enjoy Reading! Love you, Mom and Dad, 10-21-1997." *Next of Kin* is primatologist Roger Fouts's memoir about his life with a chimpanzee named Washoe. Raised as a human child by Allen and Trixie Gardner in the 1960s, Washoe becomes the first chimpanzee to learn American Sign Language; she acquires a vocabulary of over three hundred words. What I find most interesting about her use of sign is that Washoe invents her own words. All languages have mechanisms for generating new words. In ASL, compounding—combining two words to create a new word—is the primary generating mechanism, and this is exactly what Washoe does. DIRTY-GOOD is her phrase for going to the potty because it is both dirty and good. She refers to the refrigerator as OPEN FOOD DRINK, even though her human caregivers call it COLD BOX. She describes a swan as WATER BIRD. Washoe, too, is a poet.

*Next of Kin* clearly depicts nonhuman animal intelligence and personalities, something I, like my father, understood intuitively without the need for scientific experiments.

What I find most moving is the last part of Fouts's book, which explores the ethical implications of animal research. Fouts lays out the tragedy of this whole experiment—how we've robbed these beautiful beings of their family, of their wildness. It is so reassuring to find someone in the science community speaking up for animals. I am always interested in animals, but wary of the ways people study them—caged in captivity, languishing in zoos, or stressed in laboratories. What are ethical ways of knowing them?

Washoe and her chimpanzee family sacrifice so much for human knowledge. Unable to survive on their own in the wild, they could never go back to Africa. Fouts moves Washoe and her family to Ellensburg, Washington, to the Chimpanzee and Human Communication Institute. The setup there is a pseudosanctuary, with graduate students and volunteers devoted to captive chimpanzee care and their psychological well-being.

After finishing *Next of Kin*, I sign up for an American Sign Language class, so I can attend a summer program at CHCI. Books and stories about our fellow animals will become my teachers and guides, setting me off on a series of adventures, and serving as my steady companions to navigate our uncertain and often violent world. I consider *Next of Kin* to be my "Spark Book," the book that sparked the journey I am on right now—the book that I am writing to you. It prompts two questions that I carry with me today. How can we learn about, and from, other animals without harming them? How do we repair the harm we have already caused other species?

When I arrive in Ellensburg, Washington, in 1998, Washoe smacks the two thumbs of her fists together, asking to see my shoes. The young, inquisitive, at times mischievous, girl whose childhood I have read about is now in her thirties and transformed into the matriarch of her adopted chimpanzee group of five. Part of my assignment is to help with environmental enrichment in the chimpanzee enclosures, to keep the chimpanzees from getting bored in captivity. We simulate forages, placing treats in challenging locations, which require natural behaviors like climbing and using tools. Having been raised as human children, these chimps also enjoy flipping through magazines, checking themselves out in mirrors, and brushing their teeth.

Tatu, another signing chimpanzee in Washoe's family, greets me by either asking for cereal or wanting to play peekaboo. Her fingers are slender, and her signing is very polished. Black is her favorite color, and she'd flip through books and point out all the things that are black—THAT BLACK. She enjoys puzzles and foraging. But she also rocks herself back and forth, a behavior that emerged decades earlier when she was separated from her mother at an early age to enter research.

*Next of Kin* confirms what I have always believed about animals—we are all, at our core, sentient beings. Not only do we feel our own pain, but we can feel that of others. Is this understanding my inheritance from my father, and his from you?

Washoe understands this too. Scientists want to see if Washoe would pass sign language on to her offspring. Before adopting another chimpanzee, Loulis, as her son, Washoe suffers multiple losses of her biological babies. Later, one of her human caregivers, a graduate student who is pregnant, would often sign with Washoe about the baby in her belly. The woman, however, suffers a pregnancy loss. When she sees Washoe again after some time, Washoe is curious why this grad student has been away so long. The caregiver decides to tell Washoe the truth, and signs to her MY BABY DIED. Washoe replies with a finger to the eye—CRY—followed later by PLEASE PERSON HUG. Chimpanzees recognize loss across species. This, too, is a bridge.

# SOIL MECHANICS

I REMEMBER THE DAMP BASEMENT SOILS LABORATORY OF Cooper Union. We are stacking sieves, like a vertical pile of steaming dim sum baskets. We arrange the cylindrical pans with the biggest screens on top and the finest at the bottom. We pour a sample of soil on this stack and shake it. After, we separate the sieves and see what accumulates on each, differentiating the gravels from the sand, with the finer silts and clays that pass through.

I feel like a sieve writing this story, carrying only what I can collect, hold, and remember, wondering about the finer parts that slip through.

A year after the summer with the signing chimps, I move west from New York City to Berkeley, California, for graduate school. I am interested in soil behavior and how the earth responds to human pressures. It is August 1999, and a major earthquake has just devastated Turkey. My professor, Ray Seed, describes the affected area as "smelling like forty thousand dead people," further noting that "engineers who know that smell do their work a lot differently than those who don't." It is this sense of social responsibility that leads me to continue to pursue engineering, to view engineering as social work—like you do.

What Professor Guido first instills in me is that geotechnical engineering is both an art and a science. Unlike steel or concrete, the earth is not homogeneous, and our understanding of it is still new. Geotechnical engineering is about making decisions for the public good based on limited knowledge. It is about learning to work with the unknowns.

At Berkeley, I attend the lectures of Ray Seed, son of the geotechnical legend, Harry Bolton Seed. Professor Seed is a natural storyteller. He ascribes personalities to soil types to describe their behaviors—how silts have the most fun because they sometimes behave as sands and sometimes as clays. When talking about different soil instrumentation techniques, he argues for the need of multiple data sources like Standard Penetration Tests and Cone Penetrometer Tests. For the former, you collect data every five to ten feet and measure the soil's strength by how many hammer blows it takes to drive the hole a foot deeper. You can retrieve that sample of soil and touch it, rub it between your fingers. The latter is a probe that provides continuous strength data, and it graphs changes in resistance between different soil types, but you never actually see what's in the ground. Professor Seed invokes a nursery rhyme to explain why you need both: "Jack Sprat could eat no fat / His wife could eat no lean. / And so between them both, you see, / They licked the platter clean."

I wonder how many of these anecdotes are passed on in his family from parent to child. What does it feel like to grow up with access to this knowledge around you? I wonder what it would have been like to learn about engineering from you.

My numerical geomechanics professor, Juan Pestana, has so much faith in me. This subtle nurturing is something I need at this time. I love his class, applying sequences and series from calculus to approximate geotechnical models. Everything is an approximation—getting nearer toward a truth.

My engineering geology professor, Nick Sitar, takes us hiking. He tells us how to record strikes and dips of rock outcrops.

He knows how to read a landscape, how to tell if igneous rock is intrusive or extrusive. Did it form from magma deep within the crust of the earth and cool slowly? Or did it form at the surface from lava that cooled rapidly? I practice holding new geological words in my mouth: *diorite, basalt, obsidian, oxbow lake*. I am not only getting a new understanding of the earth, but of those who tried to understand its engineering principles. Geologic time is long. Geotechnical engineering is young. Its history is founded on the empiricism of a growing handful of engineers and scientists whose names are inscribed on our textbooks: Karl Terzaghi, Arthur Casagrande, Albert Mauritz Atterberg, and Ralph Peck. So much of what I learn comes from their experiments with marine sediments in US coastal cities—Boston Blue Clay, Virginia coastal clays, Chicago glacial clay, and San Francisco Bay Mud. These students and teachers of the earth compress core samples extruded from the ground to mimic in the lab what it is like in situ. They are also my forebearers, and their history is available to me.

They study how soils behave under pressure and how they fail. What can we learn from these failures? How can we be better, safer? I want engineering to be taught through an ethical or moral lens, and the faculty at Berkeley seem to take on a more caring role in ushering in the newest members of this still relatively new field. There is an ethical responsibility to impart all their knowledge in a field where so many mistakes can happen and so much depends on intuition and judgment. They are teaching us how to think and act in uncertain times.

In Professor Sitar's engineering geology reader, a purple spiral-bound collection of xeroxed journal articles, I come across one about tropical residual soils—red, lateritic, iron-rich soils that weather in place. These are the soils of Africa, South America, India, and Burma. The tropical residual soils in the lands of my ancestors behave differently than the marine coastal clays that are the foundation of my education here.

After graduating, but before my UC library card expires, I check out a biography of Karl Terzaghi, who is considered the "father of soil mechanics." I am attracted to its title, *The Engineer as Artist*, and am hoping to find inspiration and guidance in navigating this field. I don't get very far in before I stop reading. When Terzaghi first arrives in New York and is becoming the "father of soil mechanics," he impregnates a woman, and doesn't want to deal with being an actual father, so he sends her to Mexico to live with a family. I stop looking for inspiration in Terzaghi.

But, twenty years later, I return to the book. There are details in his life that still disappoint and offend. Goodman shows Terzaghi as a complicated and flawed figure who operates with some useful guiding principles. Terzaghi told his college students, "Very few people are either so dumb or dishonest that you could not learn anything from them." This echoes what a writing mentor would tell me many years later about reading: "Pan for possibility, not criticism." So, I pan. Terzaghi is working in unfamiliar terrain and writing with no models: "This book has no antecedent," he writes. I learn he is later romantically involved with the poet Sylvia Plath's mother, enjoying literature and the arts together. He then marries Ruth Doggett, a geologist. I wonder how much of his success is credited to her invisible labor. I think about Patti and Kasturba, too, and I wonder the same of them. I think about you and Gandhi, men who were legends, who I'd come to learn were also flawed in different ways. I am reminded to pan for possibility, for what I can still learn from you.

More than fifteen years after graduate school, I am at a book reading by the author Q. M. Zhang at the Asian American Writers' Workshop. She confesses that when she thinks she is nearing the truth in her work, her nose gets all stuffy. I have never thought about it before, but that resonates with me when I hear it. When reading or writing, and I feel like I'm nearing some sort of epiphany, my nose starts to tingle, and a nervous energy in my gut appears. I remember that feeling reading about tropical residual soils in Professor Sitar's geology reader. Something is triggered. Curiosity? Connection? Desire to learn about the earth from a different vantage point? There is still so much I don't know. And maybe I want to have a different relationship with the earth, than merely an engineering one. I am still craving an ethical guide and mentor.

Why didn't I learn to garden from my father? What would he have taught me about soil? What do you know about the earth that I can't find in my textbooks or courses? And what about all the women in our family, whose stories are even more invisible? Why are there no records?

Or, maybe, I was not paying attention to the ephemeral records the women—my aunties—lay out every day. Rising early each morning, their bodies bent to greet the earth. They pour fistfuls of rice flour to form flower petals, dots, squares, and other geometrical shapes—an embodied, mathematical practice. This ephemeral art will nourish ants and worms and other creatures. In her book *Feeding a Thousand Souls*, Vijaya Nagarajan learns from Tamil women that their kolams have multiple purposes: to bring beauty, invite auspiciousness, and "ask forgiveness to the earth goddess Bhudevi for our walking and stepping on her." I now practice making these patterns, not on the earth, but on our living room floor, and not with rice flour, but with dog kibble, feeding not one thousand, but one hungry soul.

# IN SITU

In my twenties, I spend a lot of time playing with dirt, or soil, to be precise. My first major field assignment is on the Calaveras Dam, about fifty miles southeast of San Francisco. I slip on my hard hat and tie the laces of my nonleather steel-toe boots. The *vegetarian shoes* label sticks out on the tongues.

"What are they made of?" drillers ask. "Vegetarians?" they joke.

I come to know the Calaveras Dam intimately. I scale all of its benches and traverse its entire length. I watch and log as samples from its depths are extracted to the surface. I make cross sections, which connect the dots between our holes.

I imagine making a cake version of the dam and draft a list of the ingredients. Cookie crumbs for the rock fill. Graham crackers for the sand. Hard candy for the gravel. Tofu cheesecake for the clay core. Pudding for that potentially liquefiable layer we were trying to delineate. And candles for each borehole location.

This dam impounds the Calaveras Reservoir, which supplements San Francisco's water supply from Yosemite Valley. At the time of construction, it is the largest earth embankment in the world. Building of the dam started in 1918, but there is a stability failure during construction, and the filled earth collapses and slides. We suspect that the debris from that event is never fully cleared out, and the current dam sits on top of its failures but doesn't mask them.

Calaveras is located on an active fault, and we are worried about what would happen during an earthquake. My seismic analyses consist of Greek letters, numbers, and charts.

"What are those graphs?" an elderly woman sitting next to me on the AC40 bus asks me one day on my way from Oakland to Berkeley, while I study for my earthquake engineering exam in graduate school. "Is that the stock market?"

I laugh, shake my head, and say no. They are acceleration time histories of earthquakes in California.

On another bus ride study session, a physics PhD student next to me inquires about diagrams of clay consolidation in my text. Back then, I am not used to talking to people on public transit. In the crowded subways of New York, I found mental solitude. Despite being brushed up against, being breathed on, or having someone "eavesread" my paper, I can find privacy. But I engage on this bus ride. I talk about clays. How, when subject to loading, they consolidate or compress as the water in the pore space between particles gets squeezed out. But what excites me about clays is their memory.

"They are like elephants," I say. They carry with them the history of their previous stresses. *Overburden* is what engineers call it—the vertical pressure on a soil from the weight of the earth above it. How a clayey soil behaves is linked to its history, and how much it consolidates is linked to this memory.

Sands behave differently. Cyclical loading, like earthquakes, is what we were concerned about in saturated granular soils. The big risk is liquefaction—losing the ground beneath our feet. The point at which the earth is no longer a stable mass but starts to flow. "When your house becomes a boat," as one of my professors says.

In the case of Calaveras Dam, I am there to help assess the safety of this earthen structure under seismic conditions. I come to know what is beneath the surface. I rub the soil between my fingers to ascertain the fineness of each grain and submerge these particles in water. Is it silt or sand? For the clays, I roll samples of them on my hand into little worms to assess their plasticity. For the gravels, I take note of the size of the particles, how angular or rounded they are, and whether they are from an alluvial source like the streambed or a colluvial fill from the slopes that surrounded us. Methodically I translate these findings into the language of field logs:

> SANDY LEAN CLAY with GRAVEL(CL), dark yellowish brown, moist. 30-35% gravel (fine to coarse, subangular to subrounded), 20-25% sand (fine to medium grained), 40-50% fines (low plasticity).

When we hit a rough patch, I suspect we hit bedrock. I finish up the log with three letters—BOH. Bottom of Hole.

A geologist from the Division of Safety of Dams visited the site randomly to check how work was progressing. I like spending time with geologists because their knowledge of the earth's history is vast, and they are often good storytellers. This one is there to inspect me. One day he sneaks a peek at my field log, just as we are finishing up a hole. I classify the bedrock as a sandstone.

"That's not sandstone!" he scolds.

"Oh," I say, fluctuating between disbelief and acceptance of my mistake.

"Oh wait, no, it is sandstone. *Tertiary* sandstone," he clarifies, but does not apologize.

I appreciate the distinction, but as an engineer, it didn't matter for our analyses whether the bedrock was a few million or a hundred million years old. All that mattered was its strength, not its history. As long as it was stable, it didn't matter where it came from.

Part of me lives my life this way then—strong and ahistorical. I am pushing forward, carrying the past with me, but not stopping to open it up. I am never quite sure if we are who we are because of, or regardless of, our backgrounds. As a child of immigrants, that tension is constant.

Drillers come to the dam with their own histories, which I am glad they share with me. Andy tells me about his experiences entering the country from Mexico. The first time it is under his cousin's name. He gets sent back. He later returns and finds work picking strawberries. Since he speaks English, he is told he has more opportunities, and so he works his way up the drill rig.

Another driller jet-sets around the world doing a specific type of sonic drilling but goes home to a town of eight people in Montana.

"What do you do in a town of eight people?" almost everyone asks.

"Well, we actually sit down and have a conversation with one another," he says.

I know I was an anomaly to them. The drill crews don't understand why I am not married, why I don't have children, wonder how I live across the country from my family, and what I want from life.

"Why engineering?" they ask. I don't tell them about you. Can someone you never met have that impact on you? Like clays, are our histories embedded in us?

# SMALL SMALL REDEMPTION

I REMEMBER THE YOUNG UNDERGRADS AT BERKELEY PROTESTING animal abuse—students who locked themselves in cages demonstrating against farm animal confinement or against the underground vivisection laboratories.

When I am the graduate student instructor for an undergraduate soil mechanics course, the professor I work for knows I previously volunteered with chimpanzees. Whenever he asks me a question about it, my face lights up and I am all too eager to explain what I have learned about these magnificent beings. About language acquisition and tool use. About what they teach their young. About their humor. About grief.

He asks me if there is anything in civil engineering that makes me light up as much as chimpanzees. That light vanquishes as I search for an answer. If I continue with engineering, what fragments of myself would remain, and which would slip through the sieve? Do I feel a greater sense of belonging in the soils laboratory, or would I rather join the animal rights activists on campus?

You left engineering for a life of activism, and perhaps that is what connects us. Do I have to choose between animals and engineering? Maybe I don't.

When I enter the workforce in San Francisco after graduating from UC Berkeley, I am looking for something more. I want to experience living minimally. Is this what you want, too, in Kallakurichi? I have found this excerpt in Gandhi's newspaper *Harijan*, where a man named Vinoba talks about taking a vow of poverty.

> We should be like water which always seeks a lower level. All water tends to seek the level of the sea, but all water does not succeed in reaching the sea. It is given only to the rivers like the Ganges to reach the sea and be merged in it. Millions of other streams water the nearest places, some of them simply water the nearest plants and trees and are lost in the earth. But they need not rue their lot therefore. The Sea to be reached is the vast humanity. We may not reach it in a day. We can not reach it by any rigid rules. All we have to do is to strive for it with the humility of water which always seeks the lower level. The process is one of eternal striving.

Is this what you are striving for, too, in Kallakurichi? I want to work in community with the other animals with whom we share our world. Is this being like water? In California, I read volunteer diaries from this chimpanzee sanctuary in Cameroon, and I take a leave of absence from work, so I can be in that forest with chimpanzees. It is my water engineering background that catches the sanctuary's interest. Like you, water becomes my entry into activism.

At that time, they have to travel twenty miles to the nearest tap to fill up on water. They have unsuccessfully drilled two wells. They ask me, "Where is the water table?" I don't know how to use a divining rod like you do. Instead, I ask a series of questions. Where is the nearest river? What is the topography? How deep were the wells that were dug? What soils did they find? Did they hit bedrock?

I start interviewing staff and former volunteers about site descriptions. I look up geology maps of the country in the library. Most of what I can find about the area are vestiges of colonialism and resource extraction—information developed to find oil and minerals, not water. I learn about seasonal climate patterns. I compile these correspondences, maps, and tidbits in a big white binder.

My father, too, does his research. He borrows a film about Cameroon from the library. "It looks like Madras," he says, giving me his support to go. My parents are initially worried about me traveling alone, particularly as a brown American woman, six months after a terrorist attack strikes New York City, and there is a backlash on people who look like us. Their hearts ache for me across the country every time they hear about an incident of a brown person being attacked out of ignorance and fear. I don't want them to worry. But I need to go. Is following your heart, this intuition, this dream, this junoon, like divining water, or seeking to be like water?

Small smalls. Small small pikins. It is a Pidgin English phrase for little children. In the spring of 2002, I have three: Emma and Niete in each arm and Gwendolyn on my back, all just shy of one year. They arrive, as I do, at the Sanaga-Yong Chimpanzee Rescue Center in the Mbargue Forest of Cameroon, about two hundred miles east of the capital, Yaoundé. I am twenty-four years old, around the same age you are when you arrive in Burma. As a volunteer civil engineer, I am tasked with providing input on site drainage and the creation of a rainwater collection system for the sanctuary, but I end up also becoming a mother to three.

Emma and I make the journey from Yaoundé there together. She, like the other chimpanzee girls, is a product of the illegal bushmeat trade where apes and monkeys, among other wild animals, are hunted and sold as a delicacy meat to an urban affluent elite locally and abroad. Their body parts—heads, hands, arms—are found in markets and on menus. Too little to be made into lucrative meat, baby chimpanzees like my small smalls are orphaned and sold as pets or otherwise kept captive.

Emma was found tied up and screaming in Yaoundé and was rescued by In Defense of Animals-Africa (IDA-Africa) with the assistance of the Cameroon Ministry of the Environment and Forestry. IDA-Africa runs the rehabilitation sanctuary and is waging a conservation campaign for the country's remaining wild apes.

Through our eleven-hour nighttime car journey to the rescue center, Emma clings to me in the back seat as we bounce along red iron-rich laterite roads and through uneasy military roadblocks. We are traveling with Dr. Sheri Speede, the American veterinarian who started this sanctuary, her nine-month-old daughter, and another colleague with the chimp organization.

Sheri, originally from the American South, has blue eyes—like Meena Aunty—and a big smile, like Uma Aunty. She is a smooth negotiator like my dad, with a gentle but stern approach.

She leaves a secure veterinary practice in the United States and has the moral conviction and tenacity to build something new—the Sanaga-Yong Chimpanzee Rescue Center. It employs a multipronged approach: rescuing and rehabilitating chimpanzees; fighting for legal protections, education, village employment, and rejuvenation.

Though I don't make the connection at the time, you and Sheri are similar. You take these big risks, leaving behind financial security, and try to use your skills for good in places aligned with your beliefs. The chimpanzee rescue center, too, is like her own Kallakurichi—a dream, a promise, a calling.

On our overnight road journey from Yaoundé to the sanctuary, Sheri talks her way through the military checkpoints along the route. We are allowed to pass, until we get to the last checkpoint.

Three young men set up a makeshift station. I do not know if they are armed, but they confiscate our vehicle papers and won't let us pass without something in exchange. First, they ask for a ride.

Sheri steps out of the vehicle to negotiate. She tells them we have no room. We are traveling with a baby and have this letter from the Ministry of the Environment to transport a chimpanzee.

"Well give us your baby, then." The men laugh.

"I'm afraid I can't do that," she says.

"Then we'll take the monkey."

It is not the right time to educate them on the difference between monkeys and apes (monkeys have tails, chimps—who are apes—don't).

"I can't give you either of the babies," she says. She tells them the only money we have is for fuel and offers a few francs to buy some beer.

And with that we resume our travels, and I tend to the baby in my arms. Emma is indeed a small small, with big big ears almost the size of her whole head. She has a thin coat of fine black hair and a tuft of coarse white hair surrounding her baby bottom. Rope burns are etched into her legs. Did she witness her mother's death? Did her mother hold Emma tight, shielding her from the gunshots to which she herself falls victim?

The only vehicles we encounter that night are large logging trucks. In the decades prior, logging companies build roads into what were previously untouched and inaccessible forests of Cameroon, opening up the wilderness to poachers. Logging crews hire commercial hunters to provide food for them and often supply the locals with guns. Logging trucks serve as conduits for transporting stolen animal bodies to other markets. Even after the trucks leave, the clear-cut tracks remain, providing pathways to now vulnerable habitats.

More taking of trees. More taking of mothers. What is a forest without her elders?

I have read about the links between logging and bushmeat, but it becomes real for me that night as I enter the forest with an orphaned chimpanzee on my lap, while giant tree trunks exit on truck beds. Emma's species, sharing six million years of coexistence with ours, can be gone within our lifetimes. Staring at these trucks full of timber I wonder just how much they are taking out of the forests. And how much is left. Do you have the same questions of the teak in Burma, where forests were felled, logs were dragged by elephants, and trees exited on ships to serve and finance a colonial empire?

Shortly after we arrive at the sanctuary, Emma is introduced to her new roommates, Niete and Gwen. They will be quarantined together for a couple of months before joining the other infants in the nursery. Niete is healing from a machete wound on her head that stands out like a part in her thick, black, fuzzy hair. Too weak to support herself, she hangs from my neck like a Velcro plush doll. Gwen is the feisty one, curious and bold, with dark, inquisitive eyes. Her baby teeth have rotted, presumably from subsisting on trash at the beach where she is found. At the sanctuary, the small smalls are bottle-fed a formula supplemented by local fruits and rice and beans. Emma picks out the rice, Gwen prefers beans, while Niete doesn't care for either. They all welcome papayas and bananas, but on days when guavas and mangos are available, they opt for those treasured treats instead.

I delight in paying attention to their preferences and how they exert their individual agency during mealtimes.

Emma and Gwen drink from baby bottles. They'd grab on with both hands, lift the bottle up, and lean their heads back. Niete hasn't mastered this skill yet, so she drinks from a bowl, but would watch Emma and Gwen with curiosity. One day, I give Niete a bottle, elevating it slightly, and she discovers the magic of the tilt. When the bottle is empty, she refuses to let go, lifting it higher and higher, applying her newfound knowledge.

Claude, the resident rooster, wakes me up every morning at sunrise. I throw on my chimp clothes, the same clothes I wore on engineering field assignments in California. Once stained with gray coastal muds and brown sands, they are now covered with red earth and papaya. The Sanaga River washes these histories away. I have left my large-scale water infrastructure projects in San Francisco to help with small-scale water supply and drainage issues in Cameroon.

Despite my graduate degree and general familiarity with modern water supply systems supplying millions of people, I am not trained in small-scale survival solutions. Do you, Thatha, also experience this shift when you switch from being an engineer working for the Empire, to providing water to villages in Kallakurichi? How do you navigate it?

Before my travels to Cameroon, I try to school myself. I brush up on groundwater hydrology, rainwater collection, and reread that tropical residual soils article. I start taking classes in French, the second language (after ASL) I learn specifically to be in the company of chimpanzees. I have grand plans of preparing more but arrive in Cameroon with only my "What (not) to order in a Paris café" French and my big white binder. What if this isn't enough?

A few years later, I will sketch part of a water tunnel on graph paper to calculate the manifold size. I always double-check my math, because I know there are different assumptions I can make, and there are different ways of arriving at a solution. As an engineer, I always wonder if there is something I can't see from where I sit. Did you have these doubts too? Perhaps it is not doubt that I am talking about. Perhaps my approach to engineering is rooted in the Jain tenet of Anekāntavāda—acknowledging the many sides and manifolds of truth.

One morning in Cameroon, I wake up to the sound of rain on the corrugated metal roof of the clinic where I sleep. I throw on my raincoat and stand outside with my waterproof Rite in the Rain notepad, and sketch the flow of water. Where does the water want to go? I shade the little streams appearing on our sanctuary site, and circle where the water ponds, and mark where it is blocked behind a barrier. Another day, I survey the site with only my wheel measuring tool, compass, and string. I calculate slopes of the rusty dusty roads, and direct where little ditches and swales should be dug. I also sketch plans for a bigger rainwater catchment system. It is an escape from crunching numbers behind cubicle walls and inhaling diesel fumes behind drill rigs in California.

Decades later, I will learn about water-dowsing chimpanzees in Uganda. A female chimpanzee named Onyofi migrates to a new group in a rainforest and is observed to dig wells for water. Scientists think she comes from a more water-scarce region and brings with her the well-digging knowledge of her community. Since her arrival, the adult females and the young chimpanzees in the new community have also started to dig wells. They scoop dirt with their hands to create a hole, and dip their heads to drink, or sponge up the water in chewed leaves. The adult males haven't started to dig yet but do not hesitate to drink from the wells created by others. Maybe I should have looked to see if the Cameroon chimps already know the way to water. Did you know of other animal water diviners in Kallakurichi?

In Cameroon, I learn how wonderful a precious drop of sun-warmed rainwater feels on my skin as I bathe right before equatorial sundown. At night, I can still feel where the small small hands latched onto my body. During the day, I watch the small smalls play and groom. Gwen likes to be tickled, panting *hee hee hee* as she giggles. During this time, they gain strength and learn to climb trees. I don't have to worry about them running away or climbing out of reach. They never venture too far from their human caregiver. Gwen is perhaps the clingiest. Never wanting to separate at the end of the day, she holds on tight, using her hands and her even more "handy" opposable toes. It's the kind of embrace that you feel long after it's done, long after you are gone. But it is her high-pitched scream and rotten-tooth grin that haunt me. On television commercials and in movies, you could hear these noises coming out of a smiling chimp. These aren't gestures of joy. It is fear that pierces my ears. Gwen has lost her mother and is terrified of being abandoned again. In the forest I'd sing to her this song called "You Got Me" by The Roots, which

I have on a music mix my partner Wan makes for me for this trip. I try to reassure her: "'If you were worried 'bout where I been or who I saw or what club I went to with my homies,' Gwen-ny, 'don't worry, you know that you got me . . .'"

I spend my nights in the veterinary clinic at the sanctuary under the protection of a mosquito net tent. I shine my flashlight to discover the location of the resident spider, look for the visiting gecko, and then climb into my sleeping bag. It is the beginning of the mild rainy season, and we are starting to experience storms in the middle of the night. The chimps sense them first, and then I'd hear their pant-hoots when the rains come. But it is the fruit bats in the tree, who drop fruit on the clinic, that disturb my sleep. When the fruit lands on the ground it has the rustling effect of footsteps approaching. But more often, I am awakened by erratic banging from above. Instead of fallen fruit, I imagine wild chimps dancing on the corrugated metal roof, and smile. One night, I notice a cobweb-covered book on the shelf of the clinic. It is a do-it-yourself medical guide, *Where There Is No Doctor*. Thankfully, there is at least a veterinarian who also serves as the village doctor. I later learn you also provide a similar role in Kallakurichi, tending to scorpion stings, snakebites, and other ailments of nearby villagers.

One of the first patients I see Dr. Speede care for is a boy, about nine years old. He has a huge lesion on his thigh, which she treats for an infection. His grandmother has brought him in; both of his parents have died of acquired immunodeficiency syndrome from the human immunodeficiency virus.

The origin of HIV/AIDS is linked to the slaughter and consumption of wild apes, which allows the benign simian version of the virus to jump species, mutate, and spread across the globe. This young boy in the clinic is also orphaned by the bushmeat trade.

A motorcycle pulls up to the sanctuary another morning bearing terrible news. "Mon enfant!" Samuel, one of the sanctuary employees, wails. He has been in the nursery with the chimps, when he learns that his son, less than a year old, has just died. CRY. PLEASE PERSON HUG. "Mes condoléances," I offer in my elementary French.

I remember my parents worrying that I will fall sick in Cameroon. But I am equipped with vitamins, emergency drugs, and insurance that would evacuate me if anything serious happens, while people around me and their children are dying, and all they have is a veterinarian. Why am I the lucky one? Why should I be rescued?

Disparity is most apparent when visiting the other temporary forest dwellers—Big Big Oil. Several kilometers away, the Chad–Cameroon pipeline is gentrifying the forest. This pipeline, supported by the World Bank and executed by ExxonMobil and its subsidiaries, channels oil from wells in Chad through Cameroon to the Atlantic Ocean, to be shipped to refineries and sold elsewhere. They promise it will be a "barrel of hope" for the people of Chad and Cameroon.

On my flight to Cameroon from Paris, I met an engineer working on the pipeline who previously worked on similar oil projects in Burma. I come to know others from the pipeline who visit the sanctuary, and we talk about water and drainage issues. On one occasion, I see their camp. A small swath of forest is cleared to set up their temporary homes, and in front of each of their trailers is a square patch of grass. Well-secured, stocked with imported food and Western amenities, Big Big Oil is isolated from the surrounding poverty.

M., an engineer for the pipeline, stops by the sanctuary to pick me up one afternoon. He is also a geotechnical engineering alum from UC Berkeley. His classmate wrote this famous paper on San

Francisco Bay Mud that I am familiar with. "I want to show you my right-of-way," he says, thinking I might enjoy seeing his work on the pipeline. I grab my camera, jump into his Jeep, and we hit the rusty red road.

When we arrive, I see the clear-cut of trees under which the oil conduit is buried. Vegetation is being planted to restore some of the disturbances, and fiber mats blanket the slopes to protect newly seeded areas from washing away. I take in the massive scale of this endeavor. During our ride back to the sanctuary, M. is beaming with pride and asks, "Isn't this great—to be driving into the sunset in the jungles of West Africa, helping to bring environmental standards to a country that doesn't have any?"

Perhaps if they weren't building an oil pipeline, no safeguards would be needed, I think, but do not say. I know the resources of the oil company assist the sanctuary from time to time. The oil camp has refrigerators, which have been used to store vaccines for chimpanzees. On one occasion, they airlift an orphan gorilla to another sanctuary. But despite these charitable acts, and the environmental erosion control measures, Big Big Oil is still disrupting wild forest habitats and impacting the small smalls within. The break in canopy could divide primate families. Despite its big big promises, Big Big Oil will not be the answer to tackling poverty and healthcare in Chad and Cameroon. The wealth, like the oil, would be flowing out of Africa. Just like the wealth of Burma and India was drained by England.

I cannot muster the appropriate response—what I imagine Washoe signing—DIRTY BAD. My big white binder serves as a reminder of what I know, when I discover how much more I need to learn. But I do know this: In this world of increasing disparity, I want to stand with the small smalls, not the big bigs. I think this is why you quit the British to move back to India.

I stare out the window of M.'s Jeep, eager to return to my girls: "If you were worried 'bout where I'd been . . . Baby, don't worry, you know that you got me."

# PILE DRIVING

A FEW MONTHS AFTER I RETURN FROM CAMEROON, I MOVE back to New York City with Wan. He gets accepted into the New York City Teaching Fellows program and I start a job at an environmental consulting firm in the suburbs of White Plains. The following summer I am on a field assignment overseeing a pile driving operation.

*In the wee hours of the morning,*
*You might see me on the 7 Train,*
*With my eyes closed,*
*And my steel toes.*

One day, I exit the subway at Main Street in Flushing, Queens. I pull my hair back into a ponytail, tuck my white hard hat under my arm, and walk toward the big inflatable raft on Fowler Avenue where I saw union workers protest. Across the street from them were protesters of the feathered variety—chickens. Further down the street was my construction site.

*Caution: Hard Hat Zone,*
*Don't forget the safety glasses,*
*And your steel toes,*
*And by the live poultry market, you may want to hold your nose.*

When I arrive in the morning, the live market is not yet open. I do not see any birds, but I hear them behind the roll-down gate. Wings fluttering. Chirping. Screaming. I smell them too. *The foul stench of live fowl on Fowler Avenue.* I move on, rushing to get to work. I couldn't face them. Not that day.

It is my birthday. My mother tells me Lord Krishna and I have the same birthday that year.

"Oh, really?" I reply. "How old is he going to be?"

"Ageless," she answers.

I turn twenty-six years old that year, the same age my father was when he arrived in the United States in 1969 with seventy-five cents in his pocket. It is my first birthday without him. We both are not ageless.

It has been three weeks since my father's death, and I have just returned to work.

*I'm stuck in Flushing for the day,*
*Because apparently what we are flushing,*
*Is being dumped out into the bay.*

New York City was building a combined sewer overflow retention facility in Queens. Like many older cities, New York has an antiquated system, sending a commingling of stormwater and toilet water to our fourteen wastewater treatment plants.

Once inside a plant, wastewater goes through several stages of treatment. First, the "floatables" get screened, trucked, and landfilled. Plastic items, used only once, will wreak havoc on the environment for eternity. The remaining fluids enter settling tanks. Then oxygen-loving bacteria feast on the organic matters. The treated water will be separated from the solids and discharged into our waterways. The sludge will be moved by a boat and may be landfilled or reused, perhaps as fertilizers.

It's an impressive system when it works. The problem is that, often when it rains, treatment plants can get overloaded and untreated sewage combined with rainwater is discharged directly into our waterways. As the old engineering saying goes: "The solution to pollution is dilution." It's not a practice we can keep up for long. Sooner or later we all have to deal with our shit.

I am there to inspect construction of a new retention facility that will capture sewage for temporary storage during storm events, which can later be treated before being emptied into Flushing Bay. Part of the facility is a pipeline that will be supported on pile foundations—446 steel legs driven into the ground until they can't be pushed anymore. The earth's strength, like ours, is measured by its resistance.

When the pipe meets its refusal criteria, seventy-two hammer blows per foot for example, we say "the pile is home." The contractors tend to cut corners, so I have to check on them. I feel like an overqualified babysitter. But I do my job. I count hammer blows.

This site, these piles, those birds, that banging is not what I want to surround myself with. With every hammer drop my heart jumps and the ground shakes. After a while I get used to it. The ear protection helps, but it doesn't protect me from what Freddy says. He is the foreman in charge of his crew, but he is not in charge of me.

I hate being yelled at. I hate having to yell. While I am not particularly attached to this assignment, I make sure the job is done right, and that means from time to time Freddy screaming in my face as I stand my ground.

Freddy and I look like an odd pair. I would say he could be my father. But he can't. He isn't.

In some ways this assignment is good. It is easy, anyway. But my mind and my body are yearning to feel something. To feel purpose, to feel alive. Aside from Freddy, the crew members are good to me. Good enough. They look out for me, make sure I am safe and treat me with a modicum of respect. My favorite is the *pile monkey*, a young guy named Bobby whose job it is to climb up the rig to position the pile beneath the hammer. Bobby is nineteen years old, seven years younger than I am at the time, but always called me "Hon." I am amused, though in a different circumstance I would have been offended. I choose my battles on this job. As long as someone is not yelling at me or lying, I let most things slide. And on most days there, "Hon" is the nicest thing anyone says to me.

Bobby is not there on my birthday. I learn that he was killed in a motorcycle accident. Though we spend so many hours together in the same place, I don't know much about him except that he was engaged to the welder Tommy's daughter. I offer Tommy my condolences, and he does the same to me for my father. We know each other only by our losses.

Unlike other field assignments I've had, pile driving isn't a conducive atmosphere to know my field crew. I don't know the names of their wives, how old their children are, or how long they've been doing this. They don't ask me what my strange name means, how come I don't eat meat, or why I am not married. Well, except for Freddy, who tells me, "You should get married, so you don't gots to do this no more."

"You're married, and you're still here," I quip back.

I know I *don't gots to do this* no more. And I have other options besides matrimony. I don't bring up my love life with the field crew. I like presenting myself as an independent woman, not reduced to my relationship with a man, or my reproductive function. I want to be seen by my thoughts and actions, my dreams and aspirations. Pile driving isn't one of them. I start plotting my escape. Critical-state soil mechanics is of interest to me. All soils, when subject to pressure, seek to find their equilibrium, their critical state—their happy state. Is water divining your happy state, Thatha? Pile driving makes me want to find mine.

But it will take some time before I get there. I don't see how this grief will ever subside. The pain is like the perpetual pounding of pile driving. I have barely been eating or seeing people since I've come back to our apartment in East Harlem after the funeral. Wan has been taking good care of me. He has a big dinner birthday party planned for me, knowing it would be good to see my friends and eat yummy vegan food even though I don't feel like celebrating my life when my father lost his. In a couple of weeks, he will convince me to go to a bed-and-breakfast upstate at a farm animal sanctuary for a long weekend, when all I still want to do is be a hermit. Labor Day is when we celebrate the anniversary of our first date in Berkeley when we were in graduate school together, and he cooked dinner for me from his *Moosewood* cookbook and asked if I wanted to watch the game. I was worried he was talking about the Berkeley-Stanford football game that I had no interest in, but he was talking about the WNBA finals, and as a NY Liberty fan, my heart warmed.

During our drive up to the sanctuary, I panic. *What if he proposes?* The anniversary/animal sanctuary combination seems too perfect. I am not ready. I am not ready to feel joy. I shouldn't have worried. He knows that. He just wants me to be reminded of the things I care about—the same things my father cared about—and see the beauty in this world, and not just experience the shit and the pounding.

The sixty-foot-long steel leg is slowly disappearing into the ground, but the pile is not yet home. We have to weld another pipe on top of this one. I receive a short respite from the constant banging. I am not supposed to look directly at the welder's torch, so I look the other way and take notice of my surroundings. Discarded coffee cups, cigarette butts, and plastic bags decorate our excavation pit. Though we are on break, a flurry of activity persists elsewhere on the construction site. Young boys are suited up to act like grown men, and grown men act like little boys. Workers light their smokes with a blowtorch.

*Four smelly things here,*
*Diesel, cigarettes, woodchips,*
*And I'm afraid, me.*

I hear backup alarms, whistles, and hollers, and I try to decipher construction hand signals that resemble the American Sign Language I know.

*Good looks like candy*
*Excavate: to dig with paws*
*T is for hot tea*

Cranes lift big, heavy things but not my spirits. Collisions of hard hats resemble missed kisses. I dream of being somewhere else.

*Montana skies*
*And cold fresh air*
*Sure beat dusty eyes*
*And dirty hair*

My crew takes a coffee break.

*Angry man shuts up*
*Pile driver gets out of crane*
*I get to relax*

I switch from counting hammer blows to haiku syllables in the margins of my field log. The days I learn to count are more interesting and sophisticated than these. When I am maybe four or five years old, my imaginary friends are numbers. I invent personalities for these digits and assign them relationships. One is a solitary man. Two is the daughter of Four. She is the heroine and Seven is the hero. They make a nice couple and fight battles against Three, the villainess, and her partner in crime, number Nine. Five is the pitcher, and Eight the catcher in games of softball. Eleven is a skinny girl who likes to run.

I see the number Six approaching in the form of our potbellied site supervisor, Yefim. A very hearty cough escapes his body, and I ask, "Are you okay?"

"Don't worry, I have my medicine," he says. He points to his pocket and pulls out a pack of Marlboros. His hearty cough fades into a hearty laugh, one that reminds me of my dad.

Yefim is from Belarus, "a very nice place before Chernobyl," he says, "but after 1986 everything turned gray." He tells me many young construction workers lost their lives in the cleanup. He, too, is carrying his losses.

This is the first conversation that day that engages me, that captures my heart and my mind. The pile driving hammer drops, and the pounding resumes. We must drive the pile home. I return to the excavation pit, and Yefim moves elsewhere on site.

We finish up for the day, and the crew quickly disappears. Although leaving this site is the moment I have been waiting for, it is also the one I am dreading. I have to walk past the live market on my way home. This time the gate is open. I hold my breath, keep my eyes on the ground averting theirs, but notice feathers on the sidewalk.

A few steps ahead of me is an elderly woman carrying a black plastic bag in each hand. Customers are still allowed to carry live birds home. In a few years, an outbreak of bird flu will require them to be slaughtered on site.

Zoonosis, the jumping of viruses from one species to the next, will punctuate my lifetime. First it is bushmeat and connections to HIV/AIDS. Then mad cow disease, a result of feeding cows to cows. They will burn the infected cows alive in Europe. Later, when I work on protecting vulnerable species in New York City's drinking watersheds, I will learn about white-nose syndrome in Indiana bats. Then there will be *the* virus that punctuates all our lives—SARS-CoV-2. Each tied to human destruction of ecosystems and exploitation of animals.

I hear chirping from the bags ahead of me, and I know it will not last much longer.

Around the corner from the live market is the parking lot of a supermarket called Western Beef. Consumers buy live animals in one and nameless, faceless, packaged, and frozen body parts in the other. One masks and distances us from the reality; the other provides no such shield. But for the chicken, I suspect, there's not much difference. Their fate is the same.

I can't get the pounding out of my head, and I wonder, in their last moments, if that's what the birds are thinking too.

# BURMESE YEARS

After my father's death, I slowly begin my own probing of history, approaching this mission like one of my field assignments. I start compiling fragments of information from various depths, times, and places to recreate what it was like in situ. I examine these recovered fragments and try to connect the dots. As an engineer, I draw conclusions based on what I know, and assess how strong they are based on how much I don't know. Years go by. What I don't know still exceeds what I do. I hear tidbits of family narratives that don't quite add up. Dates are fuzzy. Records are lost. Letters are gone. Stories contradict themselves. Searches come up empty.

Thatha, hydrology, geology, and math are languages you and I share. I think if I can find records of your public works projects, engineering drawings, anything, I can connect to you. I don't know the exact dates you were in Burma, the details of the projects you work on, or where at all you have been posted. I send numerous queries to historians and scholars who marvel at my questions but have no information, and they all think my best bet would be the India Office Records at the British Library.

Ten years after my father's death, I receive a grant to go to the British Library. It is February, and every day in London is gray and wet.

I am reminded of the words of an animated gorilla in a zoo from a film called *Creature Comforts*: "I don't like getting rained on, and I don't like being cold. And I find that here I often get rained on, and I'm often cold." I buy an umbrella in the library gift shop and register for a Reader Pass.

The vastness of the collections housed there—the legacy of colonialism—is both impressive and unsettling. History is controlled as colonized bodies are controlled, through a rigid bureaucracy. I check my coat and umbrella in the cloakroom, before entering the Asian and African Studies Reading Room with my permitted items—a pencil and laptop contained in a clear plastic bag, to be inspected upon arrival and departure. I begin navigating the archives of an empire in search of you.

I can make up to ten requests for records a day, and each request takes about seventy minutes to fill. I pore through volumes of names until I see the words *Manakkal Sundaralingam Narayanan*—M. S. Narayanan. I find you, Thatha! I make my way through your entire service record in the Public Works Department, compiling notes in a massive table, tracking changes over time. I learn that you had a bachelor's and master's in engineering, all the places you were posted, sometimes what project you were working on (lighthouses, roads, bridges), changes in your salary, and when you passed your Hindustani and Burmese language exams (1922 and 1924, respectively). It is more information than I have ever been able to find about you, certainly more than what I could find in college when I went searching for a bridge.

I, too, have a civil service record in New York City. It can tell you I worked for more than a decade on watershed protection and water infrastructure projects, but it can't tell you my hopes and fears, with whom I share my home, or what makes me feel alive.

You can't see the hours on the subway I spend writing, the string bean and tofu dish I order at my vegan lunch spot on Queens Boulevard, how I mimic the sway of the trees in Forest Park when I run in the mornings, and how my rescue pit bull Moo Cow licks the sweat off my face when I come home.

Can you recreate a life—*re-member* a body—from the knowns and the unknowns?

In 1919 at the age of twenty-three, you receive your first job as a civil engineer for the British Public Works Department in Burma, then part of British India. You are a year younger than I am when I leave for Cameroon. You are the first in your family to leave your South Indian home. In October of 1919, the records say, M. S. Narayanan serves as a temporary engineer in Toungoo, a sleepy logging town in the middle of the country, making 250 rupees a month. By February of 1920 you are promoted to assistant engineer, with a salary bump up to 300 rupees. According to revenue reports from this time, the Public Works Department in Toungoo is charged with the replacement of timber road bridges and construction work along the road to Mandalay.

Over ninety years later, I bike around this small town, stopping along the road bridges. I am in search of memories that have been buried, washed away, or built over. Giant trucks hauling tree trunks extracted from the forests pass me on the road. I am reminded of the logging trucks I saw in the forests of Cameroon, over a decade earlier.

Logging is a huge industry in your time, too. Elephants, like you, are charged with bridge and road building. One compilation of elephant stories in Burma is a book called *Elephant Bill: The Best-Selling Account of the 1920s Life in the Jungles of Burma* by Lt. Col. J. H. Williams. Williams, or "Elephant Bill," as he is called, works as a forestry assistant for the Bombay Burmah Trading

Corporation, a logging outfit, which is partly responsible for the British annexation of Burma in 1886.

Williams is tasked with working with elephants in the forested regions up north to carry and drag teak logs to the river, where they will float down to Rangoon. Depending on the density of the tree and the flow and height of the river, it could take a year or eight years for the logs to travel downstream to arrive at their destination. At the time, three-quarters of the world's teak comes from Burma. Colonialism and logging in Burma are linked and all part of a long game of growing empire. A British veteran of World War I, Elephant Bill arrives in Burma in 1920 around the same time you do. He observes everything about the elephants—their health, relationships, communication, and how they think. "The elephant knows the margin of safety to a foot, and when the log is ten feet from the edge, she refuses to haul it any closer," Elephant Bill writes in his memoir.

I like thinking of the elephants as engineers, like us, who understand the land and her limits, but what is more important to me is learning about these moments of agency, when their bodies refuse the commands of men.

Williams's elephants work during the day, but they are left to roam and feed themselves at night. The working jungle elephants have an illusion of freedom at night. Their oozies or mahouts come to fetch them in the morning to serve the empire for another day. During their evenings off, the elephants are free to mate. Sometimes these logging elephants wander off to mate with their wild counterparts. The free elephant might place his trunk on the working elephant and smell the scent of man. Can he divine her loss? When the working elephant sees the free elephant, memories of childhood may come back. Each day she toggles back and forth between two selves.

Thatha, you, too, are toggling between being a British civil servant and an Indian family man. In July of 1921, the records say, you take your first leave from work—one month long. I think this is when Patti comes to join you. You both marry ten years prior, when you are fifteen years old, and she is only eight.

Being only eight years old, Patti doesn't fully understand what is happening on her wedding day. It is tradition that the bride wears a nine-yard saree. How many times do these nine yards go around her eight-year-old body?

During the ceremony, Patti starts to cry. Someone gives her a banana. She stops crying. Amirthalingam Uncle tells me this story. How do you remember the day?

When I am eight, I declare I will never have an arranged marriage. No one has suggested I would. But I imagine how terrified I would be if I were in that situation—one that all my forebearers endured—and how trapped I would feel. A banana alone could not remedy the situation.

At the age of eighteen, Patti leaves her parents' home to finally live with you, her husband of ten years. Her father accompanies her on a ship to Chittagong, a city in what is now Bangladesh. From there, she makes the journey to Burma. What do the two of you do in that first month together? Perhaps you take her for a walk along the Kaladan River in Sittwe. Do you observe the fruit bats hanging upside down in the trees during the day, dark sacks adorning the branches? Just nearing sunset, in that golden hour, the bats wake and fill the sky.

Do you tell her all the things you've learned about Burma, what is similar and different from home? The essential Burmese phrases like *thek that lut* to indicate vegetarian, but its actual translation is closer to *free from killing life*.

Does Patti admire the sandalwood paste, thanaka, which the women apply to their faces to keep them cool in the heat? Though you have been married for a decade, you both are strangers. Over the next twenty-two years, you will have thirteen children together. (I haven't been told of any losses.) But in those first three years, it is just the two of you in Burma. Is Patti scared, excited, or both? How do your bodies come together? Does it seem natural or forced?

Amirthalingam Uncle tells me you lived where 42nd Street meets the Rangoon River, when you were posted in Rangoon's Lighthouse Division. You build a cantilever deck that looks out over the river where you and Patti play cards and watch the ships go by. It is there, along the Rangoon River, where the treasures of Burma—like the teak extracted by Williams's elephants—pass, before being shipped out to the rest of the world.

In 1926, you are stationed for a year to Myitkyina in the northern reaches of Burma in Kachin State. I learn Myitkyina means *near the big river*, referring to the Irrawaddy. The following year, you are posted a little farther south along the Irrawaddy in Katha. What I know of Katha, I learn from George Orwell. Eric Blair, the young British police officer, works in Katha from 1926 to the middle of 1927, until he contracts dengue fever and returns to England. Emma Larkin, in her book *Finding George Orwell in Burma*, argues that what Blair witnessed in Burma led to the creation of George Orwell, a writer of conscience. In his short story "Shooting an Elephant," Orwell writes, "As for the job I was doing, I hated it more bitterly than I can perhaps make clear. In a job like that you see the dirty work of Empire at close quarters." DIRTY BAD.

In his novel *Burmese Days*, based on his time in Katha, Orwell describes the racism of the British officers as they debate whether to allow an "Oriental" into their elite gentlemen's club. Orwell alludes to the fact that the Indians and Chinese public servants did all the hard work, while the Europeans were paid more to supervise them. Through *Burmese Days*, I catch a glimpse of a troubled young man no longer wanting to serve his state. It is still from the vantage point of the colonizer, reluctant as he may be. How does the racism of empire affect you and Patti, stationed in the same small town at the same time as Eric Blair? My father and his siblings all refer to your title in Burma as "chief civil engineer." The civil service records only list you as assistant engineer, never rising above this post. Do you do the work of chief while being paid the salary of assistant?

I try to imagine the things said and left unsaid between you and Patti in this remote post.

Although I am a US citizen, as a child I am never perceived as American with my brown skin and long name no one could pronounce. There are the tiny things that are said to make me feel other.

"Where are you from?" they ask.

"New York," I say.

"But what is your nationality?"

"American, I was born here."

"But where are you really from?"

This is often meant as an othering, but perhaps it is a question I am also asking myself. Where am I really from? How do you feel as an Indian working in Burma for the British? I wonder if the British treat you as they do the "Orientals" in Orwell's novel. How do the tensions between colonizer and colonized manifest? What about the tensions between the Indians and Burmese?

Do you tell Patti about these encounters or hide them, like I did from my parents when I was in school? I didn't think they could bear it. Assimilation is a form of survival, but so much is lost in the process.

When wild elephants are first caught to work in the logging industry, they are subjected to a brutal process called kheddaring to break them in. Their oozies trap, frighten, and beat them into submission. A kheddared elephant "can be immediately recognized by the terrific training scars on its legs," Elephant Bill writes. He has a disdain for this process, and prefers to train the offspring of captive elephants, who are easier to control.

Williams works with the young calves, but they, too, at first object. In her biography of Williams, *Elephant Company*, Vicki

Constantine Croke writes: "The little elephant would be showered with treats and praise from the handlers. Despite that, the calf protested—often too upset to even collect a morsel. 'For about 2 hours it struggles and kicks, then sulks and eventually takes a banana out of sheer boredom and disgust—the expression on its face can only be compared to that of a child who eventually has to accept a sweet from a bag,' Williams noted."

Williams claims his elephants are "far nearer to the wild state than any other domesticated animal." He believes they are domesticated only for eight out of the twenty-four hours in the day. Each morning the elephant and her oozie go through a ritual. The oozie first tracks her down in the jungle where she has traveled about eight miles during the night. The oozie then sings to her. "He gives her time to accept the grim fact that another day of hard work has begun for her. If he hurried her, she might rebel," Williams writes. He writes about elephants as people, not objects, yet never questions controlling them the way he does. It is a cruel compassion to understand an elephant's desires yet steer her away from them.

What is it like for an elephant to be part wild belonging to nature, and part subject belonging to empire? When does she submit and when does she rebel?

I am interested in this tension between subjugation and freedom, complacency and rebellion, and the moments when a suppressed desire rises to the surface. I wonder how your rebellion is slowly simmering during those years in Burma. You are earning a good living, able to support your family, and send your siblings to college. Is British colonialism successful because it tricks some of its subjects into believing that they are better off under its rule? But does seeing the work of empire up close also impact you in ways that cannot be justified by material wealth? What do your kheddaring scars look like? How do you, as a vegetarian, cope with being stationed in the forest with the Europeans who hunt for sport? How does it feel to help make the roads and bridges in a country only to have that infrastructure be used to remove the riches from this land?

In 1929, you are posted in Burma on foreign service under the governing body of the University of Rangoon, where student protests gain momentum earlier that decade. In March of that year, a frail-looking Indian man draped in homespun cotton and subsisting primarily on fruits arrives in Burma to raise funds for his causes— the revitalization of rural villages, the boycott of foreign cloth, and spinning cotton, or khadi, as means toward self-sufficiency. He appeals to the Indians living in Burma. "I know there are still many who laugh at this little wheel and regard this particular activity of mine as an aberration," the lawyer-turned-khadi-activist shares with the crowd in Rangoon. But Mohandas Gandhi encourages his listeners to study more deeply "the immense bearing of the spinning wheel" not only upon their own lives, but upon those of "the starving millions of India" as well. You are in the audience and hear this appeal and meet the famed satyagrahi.

I picture you listening to Gandhi's speech, like you are listening to a familiar song. Initially, you move your body only in ways prescribed to you. Suddenly, the music changes and the song morphs into a different sound. It is startling at first, but you are more surprised that your body knows what to do. You find yourself moving in ways no one said you could. No one taught you. It is as if you always knew. Dancing is a way of remembering who you are.

I pore through the transcripts of Gandhi's speeches in Burma from 1929 to find the change in the music, to pick out the surprise that makes the unfamiliar familiar again.

What is it about this spinning wheel that captivates you? And then I read this: "If you import foreign cloth, you deny yourselves the privilege and duty of working with your hands and preparing your own cloth. This is like cutting off both your hands."

I picture you as you are described to me, wearing nothing but the white cotton loongi you spun and wove yourself, making oil by crushing sesame seeds in the fists of your own hands, and carrying your divining rod from village to village to locate water wells. Your hands are never idle.

The abbreviation PR (Permitted to Resign) appears next to your name on the yearly civil list in 1934. Because these records are found only in colonial archives, the British control the narrative. You are *permitted to resign.*

But I wonder what your resignation letter—this document that signifies your shift from engineer to activist, from civil servant to freedom fighter, from subject to rebel—says. Thatha, you, like all of us, are produced mostly of water. It isn't about resigning, but rather about restoring flow—like water desiring to be undammed.

*I will not worship your god.*
*I will not follow your rules.*
*I will not eat your banana.*
*I will not drag your log.*
*I do not need your permission.*

Like a large tusker no longer willing to submit to empire, I imagine you dropping your final log on a bridge, and then walking away.

## II

# OUR KIND OF BOOK

*For my father*

I think for a while. It's hard to put into words. Gorillas are not complainers. We're dreamers, poets, philosophers, nap takers.

—KATHERINE APPLEGATE

# QUESTION MARK

I MEMORIZE THE NAMES OF THE THIRTEEN CHILDREN OF Kallakurichi like you are a prayer or a poem.

You are a family of ducks, with the eldest Parvathy Aunty leading the raft and you, the youngest, floating behind, or a procession of Buddhist monks in descending height order, with towering Sayana Uncle in front and pocket-size Uma Aunty in the back.

I picture your names cascading down the curves of a question mark, each a key to this place of dreams, jolly, and ebbing sorrow at the bend of the Gomukhi River, fed from the Kalvarayan Hills.

        Sundaresan   Meena
          Gowri        Kamala
       Parvathy       Savitri
                      Amirthalingam
                  Saraswati
              Subramaniam
          Ananthasayanam
          Uma
          Jayamangalam

        Adinarayanan

During my junior year of college, in Professor Cataldo's hydrology class, we learn how to create a topographic map. We plot all the points of similar elevation and connect them. There is so much information packed into these few lines. The distance between each contour tells you how steep the terrain is. You can predict the flow path of a raindrop from where it lands on this map. In my searching for you and Kallakurichi, I keep circling the same subjects. I realize now they are contours. The contours of satya, ahimsa, and work. The contours of truth, interspecies care, and invisible labors. There is another commonality too. Searching for what is lost, nearly lost, hidden, forgotten, invisible, but still there. If a drop of water fell on this contour map, it would flow to Kallakurichi.

Kallakurichi is an origin story, a beginning, a middle, an end, a home, a chosen ancestral land, a birthplace (yours), a death place (Thatha's), an idea, a moral philosophy, a work ethic, a revolution, a refuge, and like your cascading names, a question mark.

# FINGERPRINT OF WATER

WATER. YOUR LOVE OF WATER IS SO STRONG THAT YOU want to swim during monsoon season. During a visit to India when I am eight years old, we enter the Bay of Bengal together. A massive tide sweeps us back to shore, and we land underneath a woman's saree. *Sorry!*

It makes sense to return you back to water.

I am about to meet the river Ganga in that oldest living city of Benares, where you once sat for your college entrance exam. To die there grants you an instant ticket to enlightenment versus the karmic cycle of rebirths. In your will, you say, you want to be returned there.

Mom, Manu, and I fly from New York to Delhi in February 2004, where we meet Jaya Aunty, her husband G. S. Uncle, and Annu, Meena Aunty's daughter. We travel by car for several days with you and Lord Krishna. I carry your ashes in a box that rests on my thighs, while a statue of Lord Krishna sits on Jaya Aunty's lap. Every day she picks flowers and makes a garland for her Krishna. We start each morning around four a.m. as the early morning mist gently kisses our faces.

Several months earlier, Mom frantically calls me from the hospital, the day after your sixtieth birthday. "The doctor asked me if there was anyone else I wanted to be here," she says in a panic. It all happens too fast. One minute you felt like you were coming down with the flu, the next it seems your premonition was coming true.

"What?" I gasp. I am at work when I get the call, living in East Harlem and commuting to Westchester County for my engineering job at that environmental consulting firm. Coworkers in neighboring cubicles hear my quivering voice. Roseann, our office admin, rushes to call for a car service. A black Lincoln town car that normally shuttles company executives takes me across the Hudson River to the same hospital where we first met. When I arrive, you are sedated and can't talk. They are preparing for a dialysis procedure in the morning and waiting for your vitals to stabilize. The doctors are baffled by the severity of your condition.

The next morning, a Catholic priest and nun appear suddenly, offering their services in the waiting room. This is a bad sign. Not because we aren't Catholic, but because they anticipate what we refuse to acknowledge is a possibility. Your vitals don't stabilize. Instead, one by one, your organs fail—gallbladder, kidney, and lungs—until, ultimately, your heart stops.

Our relatives in India can't understand how American doctors are not able to save you. "How could this happen in the richest country in the world?" they wonder.

While we are all in shock, you perhaps knew death was imminent. "Narayanan is coming," Jaya Aunty tells me, are the last words you spoke to her. You called her in Bangalore from New York to tell her about a dream you had a week before your death that revealed that your god would be picking you up soon.

Thatha had a premonition, too, about his own death. He woke up and chopped up the firewood for his own cremation.

You went to Atlantic City and came home with $7,000 for Mom to put in the bank. Was this your way of preparing for your own cremation? It seems lucky, but you believed you were under the curse of Saturn's cycle. The previous few years brought a streak of bad luck for you, starting with the time you broke your femur, right after I moved out to California for graduate school. I was not there to see you in the hospital or through your recovery.

After your fall, Mom converts our dining room into a bedroom, so you don't have to go upstairs. You thought the planets were aligned against you. I dismissed such logic as unhealthy superstition. But after your death, I second-guess everything. I wish I was a better listener, a better daughter.

Several years after your death, I travel to Maharashtra with Jaya Aunty. We visit a Saturn Temple, and I think of you, wishing you relief and release. All the homes in that town are like yours in Kallakurichi—completely open, with no locks.

---

I remember those phase diagrams from soil mechanics classes. Individual grains of gravel, sand, silt, and clay are separated by voids of air or water. I draw a rectangle and divide it into three parts to represent the three phases: solids, air, and water. I determine porosity—volume of void space divided by the total volume—and the void ratio—the volume of the voids divided by the volume of soil. I can calculate water content, the ratio of weight of water to weight of soil. There is a satisfaction in knowing what is in the voids and how much space they hold, and how you need to know all these parts to understand the soil as a whole.

---

The network of fluid-filled spaces in our body between skin and organs is called interstitium. Only recently do scientists declare this an organ unto itself, the largest one by volume in the body. Previously, the process of dissection and studying dead cells under a microscope caused these spaces to collapse. But new imaging in live bodies has confirmed interstitium as an organ, and now scientists are trying to understand its importance in connecting all the parts of us.

<hr>

You die, and I am suspended in a sea of grief—a void space whose measurements I cannot calculate. Well-meaning people want to cheer me up. They tell me that in time the pain will go away. I haven't yet experienced a loss of this personal magnitude before, but I am familiar with sorrow. Was Thatha's death your first major loss, and did it ever leave you?

For me, this new void is not something I want to forget. I want to know it—what is it comprised of, how big is it? What you must remember is that the voids in soil are part of the soil itself; it isn't that they separate individual particles, they connect them. Grief, I realize, is a part of me, and also a tether. It connects me to you and the rest of our family, to what I want to be doing with my life, to the larger world and all the animals we share it with.

Perhaps, by divining grief, I am searching for truth.

<hr>

What do I want to tell you about your wake and funeral and the thirteen days after? So many people come to say goodbye to your body. We are surrounded by so much love. So many invisible labors. When you pass, Wan is stuck in New York City. Mar's dad picks him up, and later comes over and helps with carpentry and drops off some noodles. So many invisible kindnesses. Our house is filled with family, friends, cousins. I don't think I cried so hard or laughed so hard in one day.

Sitting next to me in the car on the way to Benares, now Varanasi, my cousin Annu tells me that her mother, Meena Aunty, also had a dream the same night you do—a premonition that someone is going to die. She doesn't expect it to be you, the youngest of the thirteen siblings. Annu's eyes are like yours and Meena Aunty's—the light-eyed children of the family, Patti's favorites, she tells me. Annu says that Meena Aunty told her you two were the most subject to Thatha's rage. Your siblings likely have different assessments of these rankings. Annu also says you can only believe about 50 percent of these stories. You didn't speak much to us about the violent temper of your nonviolent father, only an occasional "Appa would get so mad," and yet I already knew.

Patti dies in 1976, twenty-seven years after Thatha. It is the year after Manu is born and a year before I am. Patti is sick for a while, but holds on for the post carrying the news of her illness to reach you in America and to receive a letter from you that you are on your way.

She, too, has her own premonitions. Her eldest, your sister Parvathy, died a few months earlier, but your siblings keep the news from Patti. In her dream though, Patti sees Thatha and Parvathy in a boat, coming to rescue her floating in water.

When you, Mom, and Manu finally arrived in India, Patti was able to see her youngest son once more and hold her youngest grandson before being submerged in water, where Thatha and Parvathy were waiting for her. You, too, want to return to these holy waters with your mother.

On the way to Varanasi, we straddle the imaginary dividing line in the middle of the road and dodge oncoming traffic like a video game. We share the road with truckers, motorcyclists, bicyclists, auto-rickshaws, handcarts, pedicabs, pedestrians, and the occasional cow, camel, or rare and auspicious elephant. Yaanai.

We enter the lands depicted in *Ramayana* and *Mahabharata* and listen to Annu tell us the history of the landscape. Epic battlegrounds. Islamic architecture. British roads now topped with side stands and telephone booths sponsored by carbonated sweetened water.

We see vibrant green and yellow mustard farms with dashes of the magenta and turquoise fabric covering the women laboring in fields. Bare-bottomed children squatting. Buffalo in water. Monkeys in trees.

With this background, I listen to Annu explain the story of Ganga, daughter of Brahma, who came from the hair follicle of Shiva and descended to Earth meandering through mountains and crossing India—a celestial river with the power to cleanse one's sins and one's soul. I imagine a mystical water roller-coaster ride beginning and ending in heaven.

When we arrive at the river Ganga, Jaya Aunty is shocked. She takes in the changes since she dipped there with you and Patti several decades earlier. "Adi must be remembering full flowing water," she says.

Jaya Aunty tells me how frightened you were being submerged into this powerful river as children. That ailing body I leave in the hospital is the skinny boy in her stories swimming in this river. Your body and this river are unrecognizable from the one in her memory. While the Ganga is flowing, it is not at the height it used to be. Parts of the riverbed are dried out.

We dip our feet in the water before the final death rites ceremony.

On this trip, I buy a book, *Ganga: Common Heritage or Corporate Commodity*, by the activist Vandana Shiva. I read about Shree Veer Bhadra Mishra, a priest who is also a professor of hydraulic engineering at Benares Hindu University. "There is a struggle and turmoil inside my heart," he writes. "I want to take a holy dip. I need it to live. The day does not begin for me without the holy dip. But, at the same time, I know what is BOD [biological oxygen demand] and I know what is fecal coliform."

I ask my cousin Annu why we say *the Ganges* when everyone in India calls her *Ganga*. She speculates that the British came up with Ganges from Gangaji. *Ji* was added to give respect. We pass by a sign that says, "Ganga is the life force of Indian culture." Somewhere along the way, Ganga lost its ji.

On the steps along the riverbank, a Hindu priest guides Manu along in prayer.

*Om Bhoor Buvah Suvaha*
*Thath Savithur Varenyam*
*Bhargo Devasya Dheemahi*
*Dhiyo Yonaha Prachodayath*

The Gayatri Mantra is the only part I recognize, and the only Hindu prayer I know, which you first taught Manu, and I listened in the background, memorizing it myself. It illuminates powers of a higher being and asks for those qualities to be bestowed upon us—for light to shine in our dark moments. Did I get that right?

The rest of the ritual is unclear to me. If only you could explain it to me now.

As the oldest son, it is Manu's duty to perform the death rites. I wish there is some official role for me. In some ceremonies, the daughter makes milk and butter-laden sweets. Mom explains to the priest that I am vegan. They do not understand why I would reject the product of the holy cow. They think it's sacrilegious. "Ahimsa" is the one-word explanation Mom utters in my defense. The practice of abstaining from dairy products seems rare in India, even among the strictest of vegetarians, but the notion of ahimsa is understood, even if it typically isn't associated with rejecting milk. While I am happy to discuss the role of women or the plight of cows, I know it is not the time. We are there to honor your last wishes.

Manu is burdened by this role. It is the son's duty to light the pyre in the cremation ceremony. Back in New York, your body is incinerated at a crematorium.

"You understand this process is irreversible, right?" the funeral home manager asks as we review the waiver forms. I scowl at the absurdity of the question.

But the finality of burning your body is harder to grasp at the crematorium. Manu has to flip the switch to start the burning.

"I don't think I can do it," he says.

I understand his hesitation. I tell him what Mom told me. The day he was born, you were so happy you have someone to perform your death rites. This is your wish for him.

When your body starts to burn, it is as if all the air in my body is released into the loudest wail.

---

We board a small boat to immerse your ashes and swim once again in the same waters with you. We pass by the various ghats along the waterfront, each with their own purpose. One is full of firewood—the cremation ghat. Foreign tourists watch from a distance as families like ours mourn our losses.

Jaya Aunty stops the boat on the other side of the river where there are no formal ghats or people. We take a dip with our clothes on. I am wearing these elephant-patterned pants I bought in Cameroon, and a shirt with Om on it. The water seems murkier on the edges, so when we sail in the middle of the river, Jaya Aunty grabs my travel water bottle, fills it up with Ganga, and pours it all over her. Then, she refills the bottle and pours it all over me. We have a Ganga bath on the boat.

Along the Ganga, we stop at one of the ghats. Wet and barefoot, we climb the steps to enter these narrow, crowded alleyways filled with cows, cow dung, and persistent vendors leading to a maze of temples. Suffering from both sensory overload and numbness, I walk arm in arm with Jaya Aunty through these corridors just like she did with you when you both were young.

I read that trying to describe Varanasi is like trying to draw a map of the universe. I declare that I never want to come back, but immediately realize that I might have to one day.

Mom, though, perhaps sensing my reluctance to return, shares her own. "Just throw my ashes in the Hudson," she says. We fill my bottle again with river water and settle on bringing Ganga to New York.

# AIR

Did I ever tell you about my earliest memories? I was a couple years old when I was in Bangalore with Mom and her side of the family. She was completing her master's degree in social work and you stayed back in New York for these two years. When you came back to visit us, I didn't know who you were. You were saddened by this. But when I finally approached and hugged your leg, your heart warmed.

Around that time, I grab hold of earlobes. I like the feel of soft, fleshy lobes between my fingers. I also latch on to golden jhumka earrings, my little hands stretching the pierced holes further. Patti and Ranjani Aunty would always remind me, pointing to their ears.

One of my earliest memories from this age is a conversation, but it is only the essence of that transaction I can recall. I speak our ancestral tongue Tamil then. During this time, I am always wandering in and out of the houses on Narayanappa Layout, among Mom's parents, siblings, aunts, uncles, and cousins who all dote on me. I love going to Nana Thatha and Apeetha's place, across the street from Patti and Thatha. I remember asking Apeetha Patti about something that was puzzling me.

"What's this?" I ask in Tamil, pointing to the space around us, between us.

"Air," she explains.

"Oh," I think, satisfied with this newfound understanding.

I reach out to grab a fistful of it. I open my hand only to discover that what is in it is gone.

It is a peculiar thing, to have a memory that took place in one language and the recollection of what happened in another. Although you and Mom would speak Tamil at home to each other, my Tamil was displaced with English when we returned to the United States.

The language I once spoke with Apeetha is like what was in my hand that day. While I can still understand a fair amount of banter, when I try to use my first tongue, I open my mouth, and there is only air.

---

Or is air, uyir, *the breath of life*, what Indologist David Shulman calls "the innermost core of the Tamil being"?

I am reading Shulman's biography of the Tamil language. I am learning about the Sangam poets—the assemblage, communities, or confluences of Tamil poets, who spanned three eras. Their histories are often mixed with mythologies and the gods who live among them. Saraswati, the goddess of books, is embodied in the forty-eight phonemes of Tamil. Saraswati sits on the tip of your tongue, as Mom would say. Always brush your teeth and wash your mouth before studying, reading, or taking exams, she advises. Never read on the toilet. If you step on a book, touch the paper with your hand and then touch your eyes.

What poetry remains of these early poets is from the third Sangam of mortals, spanning a few hundred BCE to a few hundred CE. It is secular, or perhaps Jain or Buddhist. I want to ask you about them.

## AIR

What does a Sangam of poets look like? Do they travel like geese in V formation, taking turns leading the gaggle, or display like an ostentation of peacocks? Or, do they arrange themselves like a commingling of waterways, like the Triveni Sangam of Ganga, Yamuna, and the invisible Saraswati Rivers? What collective sorrows do they carry or wash away?

# WITHOUT SORROW

**K**ING ASHOKA INSPIRES YOU TO PICK MY NAME. I remember you telling me his story. When I am in middle and high school, you think I work too hard. You ask if I want you to complain to my teachers.

"No, Daddy, please don't," I say, embarrassed. It is usually the middle of the night, when I am writing some paper at the last minute. My workload is nothing compared to yours in Kallakurichi. You want something different for us, maybe something easier.

You encourage my early writing, read my bad poetry, and listen to me recite the verses you memorize in your childhood.

It is in these late-night moments that you and I really talk. We have milk and Sunshine Hydrox cookies (as Oreos had lard in them then). During those late-night chats, your mind is sometimes somewhere else, in a different world. Thinking about it now, I realize that you couldn't sleep at night. I don't know what keeps you up, and never ask.

Our conversations often wander into the philosophical. Sometimes you tell me historical tales. The one about King Ashoka is what I remember. Ashoka, you tell me, was a fierce warrior. He fell in love with a Buddhist woman, but she was a pacifist and wanted nothing to do with a warrior.

Ashoka runs into a monk one day who confronts him. "Your name is Ashoka," the monk says. "*Shoka* means sorrow. *Ashoka* means without sorrow. How could you be without sorrow when you've killed so many people?" The monk is talking about Kalinga, home of Ashoka's bloodiest battles. When you tell me this part of the story, you are almost in tears. The way you recount this question, you capture both the concerns of the monk and the remorse of the warrior.

This question spurs King Ashoka's radical transformation and change of heart. He becomes a Buddhist, and for the rest of his reign, there are no wars in India. You love this story of redemption.

You name me after Ashoka's daughter Sanghamitra, who spreads Buddhism to Sri Lanka and starts an order of female monks. Mitra is *friend*, in Sanskrit. Sangha is *community*. *Friend of the community* is the literal translation, but you translate my name as *lover of company*. The spelling you choose for my name, *Sangamithra* instead of *Sanghamitra*, includes the word sangam, *confluence*. I think *friend of confluences* is also appropriate.

After we leave you swirling in Ganga, we visit the location of the fig tree in Gaya where Buddha once sat and attained Bodhi. The base of the tree is enclosed, and fabric hangs from a ledge. A curled-up stray dog takes refuge under the heart-shaped leaves.

Ashoka's daughter Sanghamitra brought a sapling of the original Bodhi tree to Sri Lanka. Later, the parent tree is destroyed. A sapling of the daughter tree makes the pilgrimage back to her original roots and is replanted there, where that dog is attaining enlightenment. There is something here I want to say about lineages. About how they go both ways, about how we keep our deceased ancestors alive. How you used to tell me stories and now I am the one telling them to you.

The sign for FRIEND in ASL is made by interlinking the index fingers of each hand. I have a copy of the book *Teaching Sign Language to Chimpanzees*, which categorizes the vocabulary words of the chimpanzees I knew in Ellensburg, Washington. HOME, IN, and OUT were classified as locatives. I wonder about the complexity of HOME for a chimpanzee like Washoe, born in Africa, stolen for the space program, and then transferred to language research. What is home for you? You live most of your life in the United States, but is Kallakurichi always your home, even though you never go back?

The chimpanzee Tatu's column of verbs is a poem:

BITE
BLOW CHASE
CRY
GO HUG
LAUGH
OPEN PEEKABOO
SLEEP
TICKLE.

Under the category *traits*, Moja has two listed: GOOD and SORRY. Are these our only states? Do we fluctuate between feeling good and feeling sorry?

For Ashoka, SORRY is what leads to his GOOD.

In Kalinga, where the Daya River runs red with blood, Ashoka's remorse is born. Without you to tell me stories, I turn to books for history. I learn Ashoka transforms SORRY into GOOD in the form of stone. Sprinkled across Ashoka's vast empire are rock edicts, proclaiming compassion, religious tolerance, and an end to animal sacrifice. The edicts near the red river are chiseled below an elephant rock outcrop, a pachyderm Mount Rushmore. Elephants used for war also died in that battle.

I read about an elephant named Tara in a memoir by Mark Shand, who rides Tara across India over two thousand years after the battle in Kalinga. When they approach the Daya River, Tara "moved forward reluctantly, her trunk in the air, sensing, probing." She comes to a halt and "let[s] out a loud reverberating roar" which Shand believes was terror. "With her ears extended fully forward, she backed hurriedly away, then turned and fled." Did she sense this elephant and human graveyard?

I learn that Tamil poetry is categorized into akam and puram, which are sometimes translated as *love* and *war*, but I prefer how David Shulman describes them: "inness" and "outness." I wish I could ask you about the Sangam poets. You must have studied them, too, along with Tennyson and Shakespeare. They have given me Tamil words for something I am seeking to do in English—flow toward the confluence of inner and outer worlds, between personal and planetary grief—connecting memory and history, the IN and OUT contained in the elephant Tara's loud, reverberating roar.

I wonder, sometimes, if some combination of *Ashoka* and *Sangamithra* is a more befitting name for me—Shokamitra—*friend of sorrow*. I listen to a conversation with the poet Arthur Sze, who tells us that *sorrow* is depicted in Chinese with the character for autumn above the character for heart/mind.

During my childhood, I never tell you how sad I am in the fall. Autumn in the heart/mind means the end of long summer days, impending darkness, and back-to-school blues—the United States is, as one Salman Rushdie character says, "that mighty democracy of mispronunciation."

I go through versions of this scene each first day of school.

"Um . . . hmm. Wow. This is a hard one. I know I'm gonna mess this up. San—, Sang—"

The same teachers who teach me how to sound out words get flustered by eleven letters strung together. Those that dare to attempt pronunciation either invent syllables (Sang-a-mith-*i*-ra), add letters that aren't there (Sang-*ra*-mithra), or switch the placement of a letter (Sang-am-*ir*tha).

I know the drill. I am six. I am nine. I am thirteen. I raise my hand. Warmth rushes to my cheeks. "It's Sang-a-mith-ra, but you can call me Sangu."

"San-who?"

I feel their trepidation in saying my name, like it is some sort of disease, spoken with hesitancy and a question mark.

The kids who like saying my name are the kids making fun of it. They scream it and rhyme it with other things. Sangu Bangu. (*That doesn't even make sense.*) They think they are clever saying "bless you" or "you're welcome" after asking me what my name is.

They take the palm of their hand to their mouth and make sounds they are told "Indians" make. The Indians, they are told, fight with cowboys and eat with Pilgrims. The ones they dressed up as, sporting costumes constructed out of brown paper bags and construction paper.

*If Columbus was wrong, why are they still called Indians?* I wonder back then.

In school, we are not taught about their genocide, or it was not called that in our history books. We are not taught how in the Great Plains of the United States, it coincided with the bison genocide. (Colonizers also confuse the bison of North America with the buffalo of India.)

Only as an adult did I see that chilling picture of a mountain of bison skulls. A sight that would have caused you to weep instantly.

~~~

Closer to home, so many place names in our own Rockland County have Indigenous origins from the Munsee Lenape, but their histories were not foregrounded in my classes. The children's name teasing conveys a legacy of cruelty, an inheritance of fear of other, and a country that has not reckoned with its violent past, a land built on sorrow. There is no acknowledgment of the ways that historical violence seeps into the present.

I do not have words for this then. My body holds an unvoiced fear of the simple act of taking attendance.

You ask me once if I want to change my name officially to Sangu, to make it easier "in America." I immediately say no. I can't imagine how hard it is for you to even ask that—to offer to take back the most beautiful thing you have given me, the name you pick before even the possibility of me.

I wish I knew these lines of Warsan Shire's poem, "The Birth Name," to recite to you back then.

> give your daughters names that command the full use of tongue. [. . .] my name doesn't allow me to trust anyone that cannot pronounce it right.

Or I could have had you listen to Assétou Xango's spoken word that Shire inspired.

> I want a name only the brave can say
> a name that only fits right in the mouth of those who love me right

I'm generous in accepting vowel variations in the pronunciations of my name, as long as it is said with care. I want to tell you now that I love my name and to thank you for giving it to me. Do not worry about me, because I have also come to love the autumn, with its orange leaves. Orange is the color of wisdom, you used to say. I am asked if I have a favorite season for writing. I say the fall, because that is when what I feel on the inside matches what I feel on the outside. The confluence of akam and puram. Autumn in the heart/mind.

How can you be without sorrow? I am reading about the Bhakti poet Saint Ravidas who envisioned Begumpura—the city without sorrow. The scholar and activist Gail Omvedt contextualizes the anticaste writings of the Bhakti poets as utopias envisioned not by the societal elites, but ones that imagined a society without elites. Ravidas imagined "a place with no pain, no taxes or cares, no one owns property there." In a place with no hierarchy of caste, Ravidas is transformed—"a tanner now set free."

"Utopias, the posing of alternatives, remain a crucial aspect of any struggle," Omvedt writes. "They are, in other terms, the part of social movement discourses or frames that inspire people to action by uniting ideals with an analysis that makes a claim to possible realization. They unite ecstasy and reason, projecting a future that is achievable by present action."

Is Kallakurichi an attempt at a sorrowless utopia? Or is sorrow key to a utopic transformation?

---

One morning over breakfast, Wan and I are talking about the Korean concept called han. I am scrambling tofu with vegetables and toasting bagels.

"How would you define *han*?" I ask, as I assemble a breakfast sandwich topped with kimchi.

"The untranslatable, undefinable *han*?" he replies.

"Is it like a deep grief or sorrow?" I persist. "A chamber of exquisite sadness?"

"Your jam?" he kids, acknowledging my kinship with sorrow.

Wan tells me that han lives in the body. It is felt in the body. It is the sorrow and grief of Korea being invaded, colonized, and then vivisected. Families and a nation severed.

Does this also remind you of India, or the United States? Do you know of a shared word for this collective bodily sorrow that holds the outside and the inside, akam and puram, *love and war*? Is

Ashoka's transformation to peace and ahimsa rooted in this kind of grief and reckoning?

I think about this sorrow that anchors a body, the sorrowless Ashoka, and the concept of ahimsa. I also wonder if there is a word for sorrow that extends beyond a human lens. The sorrow for animals and plants?

Do you remember that paper I was writing in middle school about deforestation of the rainforests? In one of our late-night chats, you suggest a sentence at the end: "If man continues to exploit nature for temporary gain, the permanent loss of all life on this planet including humans is sure to follow." It is a sentence I will rework in different forms for many years to come, shaping most of my life's quests.

I return to the monk's question to Ashoka and the dualisms it contained.

*How can you be without sorrow?* Is this a question of reckoning for historical harm?

*How can you be without sorrow?* Is this an aspiration to cause no further harm?

# TRUTH

REMEMBER WHEN I BECOME THE PRESIDENT OF STUDENTS Against Animal Violations and Exploitation (SAAVE), my high school's animal rights club, the one that Gina Ingaglio started?

We organize an annual fur protest in Nanuet and stencil fish drawings on catch basins to remind people that any waste drains into streams. We petition Warner Brothers to liberate the star of their movie *Free Willy*, a captive orca named Keiko, who in real life had not been freed. One year, we host an environmental summit, where outside groups are invited to set up tables in the main foyer of the school. It is the mid-1990s, and I invite Vice President Al Gore. Ms. Wilson, our group's advisor, is so excited to tell me I have mail from the White House.

Mr. Gore sends his regrets.

National animal rights groups mail me pamphlets and VHS tapes. I remember showing the footage of animals being tortured for food in Ms. Wilson's classroom after school one day. Ms. Wilson and the club members are not prepared for the videos. There is a disconnect between caring about animals (which everyone seems to do) and eating them (which everyone seems to do). To watch such videos requires sitting with the discomfort of violence, and reckoning with and reconciling that gap (which is much harder to do).

For me, that's when I begin to challenge consuming dairy. You are confused at first. You tell me milk doesn't harm the cow. You believe they only take what is left over after the calf has finished. Maybe this was true in Kallakurichi. "This isn't India," I tell you, not knowing yet that the fate of cows in India isn't any better.

You support my other early attempts at protesting animal cruelty and help me get out of dissection. In biology class, I boycott the fetal pig dissections, seeing no value in killing a pregnant sow so that teenagers could slice her unborn to study life.

"Where do you think your breakfast comes from?" our student teacher asks.

"Not from a dead pig," I say.

Turns out hers doesn't either. She is a vegetarian, too, but as a biology teacher in training, she is committed to the curriculum and sees it as no worse than other people's breakfast.

There is an anti-whaling poster in that classroom that reads *Save the Whales, Boycott the Japanese*. Animal violence is all around us in the United States, and I don't understand why the poster does not say *Boycott the Americans*, or *Boycott Ninth-Grade Biology*.

---

Later that year, we study genetics. A dent, a dip, a valley, and inheritance. Chin clefts are rare but dominant traits. Manu and I have dimple chins like you and John Travolta.

The Punnett square shows me why we looked alike, but the lessons around it reveal the other invisible inheritances from you.

I become an advocate for fruit flies. That drosophila genetic experiment seems so messed up to me—mating winged and wingless fruit flies to see how their offspring would turn out and whether they match our Punnett square predictions. After the experiment, the flies are sent to "the morgue," a sealed glass jar where they suffocate to death.

I write an elegy on a sheet of loose leaf, folding it like a card and placing it next to the glass-jar morgue. Why do humans think themselves superior and separate from other living beings? How are lives so easily discarded? In high school, I am so lonely in this questioning. Even if knowledge once came from experiments like this, do we have to continue these practices now? Are there other ways of knowing, to obtain knowledge, without harm?

I do not want satya at the expense of ahimsa. I do not want violence to be the path to truth.

～～

When I am sixteen-going-on-seventeen, I receive a scholarship to spend a summer studying Spanish in Madrid. I am good at reading the language, but shy when speaking, and hide from attention in class. But after I get a perfect score on the first exam, the teacher starts calling on me all the time. When the program takes the students to see a bull fight, I say I'm not going. The teacher asks me to explain, en español. ¿Por qué, Sangamithra? Can I find the words in this other tongue? "No es justo," I say.

On a field trip later that summer, I am on a bus to Grenada. A bee enters. The other girls start screaming, covering their faces with their long hair. One of the guys stomps on the bee. The bus lights up in cheer. You know the cheer, the same one that fuels bullies who squash who they fear and don't understand.

"Oh no!" I scream, interrupting their celebration.

～～

When I move to New York City in 1995 to study engineering at Cooper Union, it is liberating to me for many reasons. Many of my classmates are also children of immigrants, or immigrants themselves.

My unique name is no longer an anomaly. What I love about the East Village in the '90s is that I believe you can be different in all kinds of wonderful ways and feel welcomed. And when it comes to eating, I can easily find vegan and vegetarian food, and I never have to explain what these words mean. I say goodbye to meals of iceberg lettuce salads and fries when I eat out and begin a series of adventures on my student budget based on two lovely words: *Lunch Special*. There is the Burmese lunch special on East 7th Street at Mingala where all the physics teachers eat in the back. And you remember the row of Indian restaurants on East 6th Street adorned with chili pepper lights? My favorite is Sonar Gaon, where we can get a five-course lunch for $2.99. Freshman year, I'd go with classmates on Fridays before our three-hour chemistry lab. I devour samosas, mulligatawny soup, rice, roti, dhal, and alu chole grams.

On my walk to classes, I pass the "Mean People Suck" guy, the man who pushed stickers with that phrase to all the passersby in Cooper Square. I hear another woman roar: "ANIMAL RIGHTS. Sign the pe-TI-tion." There is an odd cadence to this. A syllable wrongly accented. I recognize the loneliness in her voice.

My engineering coursework does not involve animal research, and there isn't even an animal lab in the college, but I am soon confronted again with ethical dilemmas during a summer engineering fellowship at the Hospital for Joint Diseases after my sophomore year. I am contemplating exploring biomedical engineering and applying my civil engineering knowledge to the structure of the body. I am measuring acetabular cup rotation in hip replacements with my lab partner, who looks like an Indian Julia Roberts with her curly hair and signature plump lips. This assignment is all geometry and thankfully no animals are required. However, we are surrounded by animal research. Dogs have limbs broken so surgeons can practice new techniques. Rabbits are killed for cartilage research. I read about other research on the compressional strength of sheep legs, mimicking engineering experiments on steel and concrete cylinders.

One day, my partner and I are assigned to assist a handsome surgeon, who is preparing for dog surgery. This is a perk of the program, and I am supposed to feel grateful. Instead, I have so many questions: Why dogs? How do you break their bones? What happens after the surgery? The doctor is someone accustomed to charm working on his behalf. He grows increasingly uncomfortable with my questions, and my resistance to his charm. It isn't meant to be a personal attack on him so much as a curiosity about how we accept this kind of violence in the name of science. He rattles off some greater good justification, takes my Julia Roberts partner in for the surgery, and leaves me at his desk with an assignment to type up a bibliography—my punishment for taking up space and asking these questions. I am relieved to not watch his surgery, but I resent compiling his bibliography.

I haven't yet heard of Anna Kingsford. Years later, I would be introduced to her in Gandhi's autobiography, and I would seek out her writings. Kingsford studies medicine in Paris in the late nineteenth century, becoming an advocate for women and animals. She writes about her first exposure to vivisection in an article for *The Heretic*.

> Very shortly after my entry as a student at the Paris Faculté, and when as yet I was new to the horrors of the vivisectional method, I was one morning, while studying alone in the Natural History Museum, suddenly disturbed by a frightful burst of screams, of a character more distressing than words can convey, proceeding from some chamber on another side of the building. I called the porter in charge of the museum and asked him what it meant. He replied with a grin, "It is only the dogs being vivisected in M. Béclard's laboratory." I expressed my horror; and he retorted, scrutinizing me with surprise and amusement—for he could never before have heard a student speak of vivisection in such terms—"Que

voulez-vous? C'est pour la science." Therewith he left me, and I sat down alone and listened.

Much as I had heard and said, and even written, before that day about vivisection, I found myself then for the first time in its actual presence, and there swept over me a wave of such extreme mental anguish that my heart stood still under it. [. . .] And then and there, burying my face in my hands, with tears of agony I prayed for strength and courage to labour effectually for the abolition of so vile a wrong, and to do at least what one heart and one voice might to root this curse of torture from the land.

Kingsford and I have this twinning experience. I did not have her writings for solidarity back then, and perhaps knowing we are still fighting the same battles over a century later, she may not have provided much comfort, but at least some validation. But I am drawn to Roger Fouts's book about Washoe, which you and Mom mail in a care package the semester after this fellowship. Like Kingsford, Fouts's ethical awakenings about the use of animals in research is something I welcome to hear from the scientific community.

I don't know it at the time, but several blocks away from my apartment on Fourth Avenue is *Satya* magazine's office at that time. Fouts was featured in *Satya* back then, as well as the fight to liberate laboratory animals at NYU. In November of 1997, sixteen members of NYU's Students for Education and Animal Liberation occupy their college president's office, demanding that the remaining chimpanzees of NYU's Laboratory for Experimental Medicine and Surgery in Primates (LEMSIP) be transferred to sanctuaries rather than to another laboratory notorious for its animal welfare violations. After twenty hours of occupation, NYU agrees to the protesters' request to retire the remaining chimps. Through the physical act of placing their bodies in the highest office on campus, the students reveal concealed truths and ultimately free chimpanzees who I will later meet in sanctuaries in Montreal and in Texas in the years to come.

If I think now about my East Village in the '90s, of course *Satya* was always there—in the laminated restaurant review posted on the window of my frequented veg joints, or in a large stack on a floor with the new age journals at the health food store. We are inhabiting the same city, struggling with similar ethical quandaries, and not knowing each other, yet.

∼∼

"It's a terrific magazine, full of depressing shit," Nellie McKay, a vegan folksinger and songwriter, says about *Satya*.

I never get the chance to tell you about it, so I am telling you about it now.

It comes into my life in the summer of 2003, when you leave it. It becomes a home for my grief that is personal and planetary.

Wan and I decide to attend our first animal rights conference in DC, and volunteer to set up a table to raise funds for the chimpanzee sanctuary I volunteered for in Cameroon. I display pamphlets and chimp T-shirts and a laptop that plays a video slideshow of my small smalls. I purchase an extra-large cream-colored chimp T-shirt to give to you. I have a similar one in green that I wear at the sanctuary. It has the face of a chimp on the back. Gwen, one of my small smalls, picks at the face with her fingers, trying to groom the image.

You leave us two days after your Kallakurichi Birthday that year, one month before your official birthday, and I don't get you the shirt in time. We place it in the wooden coffin with you before you turn to ash.

I don't know our time together is so limited when I attend this conference. Our table is right next to a Brooklyn-based activist magazine that focuses on all the issues near and dear to us. It looks familiar to me, and I recalled seeing it at restaurants and health food stores with other free literature but had never picked it up before. When I do now, I feel so at home and instantly connected.

Wan and I meet editors Catherine Clyne and Rachel Cernansky, and I discover such thoughtful writing that uncovers the connections between people, animals, and the planet. We make new vegan friends who live in our city.

After we return to New York, I spend my lunch hours at work reading back issues of *Satya* and learning about its history. In June 1994, Beth Gould and Martin Rowe, who met as graduate students at New York University's Religious Studies program, decide to start a community paper.

Thirteen years later, Beth reflected, "I don't think we were naïve enough to think, at its inception, that one magazine, with limited circulation and an even smaller budget would single handedly bring about a vegetarian revolution. Instead, we sought to provide a home for those who struggled to make the world a better place using the tools of education and compassion, rather than anger, recrimination and rhetoric. The name we chose, and the history of its antecedent, satyagraha, inspired our intention to change minds using nonviolence."

Beth and Martin first plotted *Satya* together at a now-defunct vegetarian restaurant. Copies of *Satya* are distributed free of charge in veg restaurants and other community centers throughout the city. It is a beautiful leap of faith—sending papers out into the world imagining they will find their way to willing readers.

I wish they could have found their way to you.

*Satya* is more than just a publication. What unfolds over the years is an evolving conversation around ethics, justice, and activism. These issues are both local and universally bound. *Satya* builds a community that takes root in NYC and expands globally, reaching readers in six continents.

"Animal advocacy, vegetarianism, and environmentalism are ideas whose time has come," Martin declares in that introductory issue. "The issues they raise are surfacing more and more in the public's mind as both individuals and groups of people realize that the quality of their lives and the lives of those they love is being diminished by rampant greed, exploitation, toxicity, and thoughtlessness."

In that first year of publication, this expression, *an idea whose time has come,* frequents the pages of *Satya*. It comes up in reference to the creation of a vegetarian resource center in NYC, advocacy for the ideas of animal rights and environmentalism in general, and the magazine itself, which takes an intersectional approach to addressing these issues. Do you think it is an idea that has already come? These are the governing principles of your childhood in Kallakurichi, and I have become obsessed with tracing their lineages.

*Satya* provides a forum for those who are and want to be activists. In the early years, there is even a periodic column titled How to Be an Activist. In the later years, the column disappears, but the question is implicit throughout the pages. The voice is not so much prescriptive but inquisitive. What do we do with what we know? How do we invoke change in our society and in ourselves? And how do we deal with that most pernicious and dangerous question: What's the point?

Uncertain of the outcomes, activists commit to change. "What might happen is what happened to me," Martin writes in an editorial in 1996. "Progressively, an everyday sense of cynical powerlessness in the face of violence on the street and the abattoir, in the war zone and in the forest, is transformed into a sense that at all times, I can choose not to be powerless by removing as much violence as possible from my life and leaving myself open to the absurd possibility of actually making a difference."

Soon after I find *Satya*, I am mourning you. I wish I knew back then the words of the psychotherapist Francis Weller, who said, "Grief is subversive, undermining the quiet agreement to behave and be in control of our emotions. It is an act of protest that declares our refusal to live numb and small."

I do not want to live numb and small. Losing you, I gain a certain courage.

I quit my pile driving job to join *Satya* full-time as the assistant editor under the leadership of Catherine Clyne, whose writings, intellect, and moral clarity I greatly admired. In one editorial, Cat details the anatomy of her own awakening: "My vegan, animal and environmental awareness was a slow process of connecting the dots that continues to this day. It began with an inarticulate gut-feeling, stemming mostly from experience, rather than bibliography."

I, too, was slowly connecting my own dots. I think you would appreciate me quitting engineering to become a writer like you, and joining a magazine inspired by the same satyagraha movement that inspired Thatha. I want to channel Thatha's conviction and your compassion to navigate this world that is often cruel. *Satya* profiles global and local activists and hosts debates and discussions over tactics and approaches. What is the linkage between individual action and systemic change?

I work at the magazine during the rest of the George W. Bush presidency (yes, he was reelected for a second term). I can't separate the light I found in working on its pages from my awakenings to the darkness of those times. The intentional, invisible violences. The "I can't recalls" and the lack of accountability. The normalization of torture and waterboarding. The Orwellian erosion of truth—the "clear skies" initiatives. The criminalization of activism. Truth was eroding around us.

# THE CHICKEN ISSUE

When Nellie McKay calls *Satya* a terrific magazine full of depressing shit, she is reading our issue about Hurricane Katrina. The storm devastates the Gulf Coast, and *Satya* covers the racial and environmental injustices the storm exposes, but also the harms to animals. One of our contributors, Miyun Park, writes how one observer described the dead chickens looking like a "field of cotton." By the time Miyun gets to the site to attempt a rescue, "Tyson's crew is already on-site, bulldozing the remaining sheds and whatever chickens were alive inside." The artist Sue Coe creates an original painting for this issue, to accompany an article on the laboratory primates left to drown in a university basement, "Help Was Never on the Way."

After Katrina, *Satya* is responding to distress signals closer to home.

One Sunday in October, I receive an unexpected phone call from our managing editor, Kym.

"You want to help transport a few hundred chickens to the Catskills, tomorrow?"

Kym and I have been working together for about a year at this point, and she has volumes of animal rescue stories. She grew up raising dozens of animals and tending to injured wildlife. She shares her Brooklyn apartment with her sister Kristi, three cats, and three dogs, one of them a recently rescued, elderly, senile, blind Maltese named Jack, who is almost always attached to her. He would bite himself till he bled if left alone. She carries him in a bag, brings him on dates to bars and restaurants, and he sleeps under her desk at work, except when he decides to pee in the middle of the office.

Kym is the person I go to when I have any sort of question about caring for animals. When Wan and I are about to move out of our East Harlem apartment, I tell her about one of the feral cats in the backyard. He's orange and we call him Saffron. I want to try to trap, neuter, and release him before moving out. Kym lends me her trap. She knows Wan is allergic to cats, but says that if we trap the cat, we need to keep him in the apartment overnight to make sure he doesn't eat before the operation.

"Can you put him in the bathroom for the night?" Kym asks.

"Wan or the cat?" I joke.

Her question to me is one I hear many times over the years. Someone would find an animal in need of a home, and Kym would ask, "Can you keep her in your bathroom? Just for the night?" And if I can't, the animal ends up in hers. Cats, dogs, roosters, and ducks have all left their paw- and footprints in her tub.

At the time, her younger sister Kristi is a humane law enforcer for the ASPCA. "Kristi was called in on a case in Brooklyn," Kym tells me. "There was a makeshift market of chickens, left out on the street without food or water." Neighbors complain about the stench.

The problem is too big for Kym's bathroom.

Kym and Kristi take it upon themselves to find homes for them.

Jenny Brown of Woodstock Farm Animal Sanctuary and Kathy Stevens of Catskill Animal Sanctuary volunteer their services to the initial triage. They take in some of the birds and provide temporary shelter to the rest, as other homes are located. Kym calls to ask for my help in the transport and triage.

What you must understand is that *Satya* is more than a magazine, more than a job. The boundaries between work, life, and activism are all blurred.

"Yeah sure, I can go tomorrow," I say.

I agree to go but am not sure what I can contribute. By this point, I know a fair amount about taking care of chimpanzees, but I don't know much about chickens. But Kym needs a backup driver, in case she has to ride in the chicken van with her sister. I can drive and take orders as needed.

"Just tell me what to do," I say, thinking of the live market I used to pass by in Flushing, on my way to my pile driving job two years prior, right after you passed away. I no longer want to avert my gaze.

I meet Kym, Kristi, and Cat the next morning at the ASPCA on 92nd Street. The chickens were moved from the street to the basement there. They spent the night in the same crates in which they were abandoned. The rains from the week before drowned some of the birds on the lower crates. Kristi and her partner separated the dead from the living. At *Satya*, we see the impact from the rains of Hurricane Katrina to the rains of Brooklyn on the lives of chickens.

Having been stacked on top of one another for days, the Brooklyn chickens are urine-stained and feces-crusted. We load the crates of chirping yellow feathers into the van.

On the way up we drop Jack, Kym's dog, at her parents' house for them to babysit. Only on rare occasions is Jack away from Kym, and this chicken rescue makes the cut.

When we arrive at the sanctuary, a small group of volunteers gather to set up a medical check station in the barn where the chickens would be temporarily housed. We unload the chickens one by one from the crates. Jenny Brown and Kym put ID bands on each bird, perform basic medical checks, and fill out individual medical record forms. They mend wounded toes and sprained wings. They gently wipe blood and excrement off of feathers.

The birds in the crates are anxious, and we decide it is better to just let them all out. We corral off a section of the barn and unload all the birds who haven't been checked and tagged yet. On the other side, we give food and water to the birds after their exam.

I am assigned the task of being a chicken catcher and pick up birds who need to be examined and hand them to Kym and Jenny. I am hesitant at first. I don't know how to pick up a chicken. I cautiously approach, but I am too slow, and their wings flutter. I let go. The last thing I want is to scare them. Kym shows me how to place my hands quickly and gently over their wings and swoop them up.

After their checkup, we set them free in the barn. Food and water stations are set up.

"Make sure they don't drink or eat too fast," Kym tells me.

Once outside, their chicken behaviors start to show. They rub their feathers on the ground and take dust baths. They nest.

One chicken's eye is covered with feces. Another volunteer wipes it clean, and the bird can see her new surroundings for the first time. She starts exploring.

There are only a few hundred of them in a well-ventilated barn, but the dust and ammonia from the urine-soaked feathers begin to assault our lungs, and we put on masks.

"I think I have bird flu," Kym says, half joking.

"Me too," I say.

When I am done catching chickens, I walk around the barn monitoring their conditions, tasked with looking for critical birds.

"What should I look for?" I ask.

"You'll know."

I find some birds that I thought were immobile, but it turned out were just sleeping. But there are a few who are particularly weak and unable to support themselves. A makeshift chicken ICU is set up. Cat and Jenny administer IV fluids. They sit on the floor of the barn with the critical chickens bundled up in blankets on their laps. Most of the half-dozen birds in the ICU will not survive the night. But I watch Cat and Jenny care and fight for those tiny lives bred for slaughter and sacrifice and discarded as garbage. They offer comfort and tenderness in their dying moments.

As for the rest, they will find new homes to live out their lives as chickens. Kym and Kristi will convert a U-Haul into a makeshift chicken transport vehicle, dropping them off in sanctuaries across the East Coast. This chicken episode—half the office taking the day off from work to rescue chickens—becomes the work. It inspires an entire magazine issue devoted to the most eaten, least protected, species.

We drive back to Brooklyn that night mostly in silence. Our clothes are stained, our hair stinks, and our nostrils are stuffed. We pick up Jack and he is excited to be reunited with Kym. He sniffs our strange smells and howls in the front seat. *Where have you been?*

# A THING AFTER YOUR OWN HEART

When I read Gandhi's autobiography, I want to talk to you about the London section, about what it was like to be an Indian and a vegetarian in the heart of a meaty empire.

In 1888, Gandhi travels from India to London to study law. He arrives carrying a promise to his mother that he will not touch meat (or wine or women) while in Europe. At the age of nineteen, Gandhi is both a reluctant vegetarian and a reluctant meat-eater. Many of his Indian peers take to meat-eating. As one friend explains to him, "We are a weak people because we do not eat meat. The English are able to rule over us, because they are meat-eaters. You know how hardy I am, and how great a runner too. It is because I am a meat-eater. Meat-eaters do not have boils or tumors, and even if they sometimes happen to have any, these heal quickly."

Gandhi recalls this doggerel of the Gujarati poet Narmad that his classmates recite. Have you heard this one too?

Behold the mighty Englishman!
He rules the Indian small,
Because being a meat-eater
He is five cubits tall.

"It began to grow on me that meat-eating was good, that it would make me strong and daring, and that, if the whole country took to meat-eating, the English could be overcome," Gandhi recounts in his autobiography.

His first experiment with this truth is traumatic:

> We went in search of a lonely spot by the river, and there I saw, for the first time in my life—meat. [. . .] The goat's meat was as tough as leather. I simply could not eat it. [. . .]
>
> I had a very bad night afterwards. A horrible nightmare haunted me. Every time I dropped off to sleep it would seem as though a live goat were bleating inside me, and I would jump up full of remorse. But then I would remind myself that meat-eating was a duty and so become more cheerful.

I imagine you might have similar nightmares just reading about this incident.

Gandhi attempts to eat meat for another year, while hiding it from his parents, but becomes too conflicted.

And so, he starts his journey to London with a vow to be vegetarian.

"It is all very well so far but you will have to revise your decision in the Bay of Biscay," an English passenger on the ship warns him. "And it is so cold in England that one cannot possibly live there without meat."

"But I have heard that people can live there without eating meat," Gandhi replies.

"Rest assured it is a fib. No one, to my knowledge lives there without being a meat-eater," his shipmate continues.

After he arrives, the young Gandhi struggles, subsisting on porridge and too few slices of bread. (He is too shy to ask for more.) One evening he stumbles upon the Central Vegetarian Restaurant on Farrington Street. "The sight of it filled me with the same joy that a child feels on getting a thing after his own heart," Gandhi writes.

This line reminds me how I felt when I first encounter *Satya*.

At this vegetarian restaurant Gandhi consumes his "first hearty meal" since his arrival in England (though he doesn't tell us inquiring minds what it was), and discovers an essay that changes his life.

In the display case of the restaurant is a copy of animal rights pioneer Henry Stephens Salt's pamphlet *A Plea for Vegetarianism*, which Gandhi purchases for a shilling and devours with as much delight as his meal. It is in this moment that Gandhi recommits to vegetarianism by choice, and not out of familial duty. He later comes to believe ahimsa and vegetarianism are key strengths—not weaknesses—in the effort to fight for Indian independence.

I read Salt's *A Plea for Vegetarianism* in a vegan restaurant on Queens Boulevard more than a century after Gandhi. I devour it with perhaps as much delight, remarking on Salt's wit and brilliance, and how his writings remain relevant today.

Salt begins with a confession—that he is a vegetarian—and that "this is rather a formidable admission to make, for a Vegetarian is still regarded, in ordinary society, as little better than a madman, and may consider himself lucky if he has no worse epithets applied to him than humanitarian, sentimentalist, crotchet-monger, fanatic, and the like."

He shares how most people don't understand how one could not eat meat, while others admit the possibility of a meat-free existence, "but profess themselves, with a pitying smile of superior intelligence, utterly unable to imagine any reason for such abstinence." Salt lays out the arguments for vegetarianism (economic, scientific, and ethical), but also laments the pushback from skeptics, often lodged in the form of a never-ending list of dismissive questions. "In truth, it is but a thankless task to answer them at any time, for they are hydra-headed monsters, and spring up as fast as one can cut them down," Salt writes.

By laying out these challenges associated with advocating for animals where cruelty is the norm, Salt's writings offer solidarity to those questioning this kind of societal violence and become a call for action—for at least one significant reader—rather than an excuse to retreat.

Such a moment inspires the soon-to-be activist Gandhi to cut his teeth in writing and organizing for the London Vegetarian Society. This continues to shape his thinking and provides a foundation for future activism.

I visit those early articles Gandhi wrote in *The Vegetarian*. They read as "India explainers" to me. I would distill them like this: *Dear English people: Let me describe India to you. Let me describe Indian festivals. Let me tell you about Indian vegetarians. Let me remind you that not all Indians are Hindus, not all Hindus are vegetarian, but that most Indian nonvegetarians are vegetarian most of the time. Let me tell you the difference between the Indian meat-eater and the English meat-eater—the Indian meat-eater does not think he will die if he doesn't eat meat.*

Some of his fellow English vegetarian friends and readers acknowledged their Indian influences, having been raised in the colonies. But Gandhi soon begins to trouble this relationship with empire, that I would paraphrase as this: *Let me describe foods of India to you, you who colonized our lands for teas and spices, but did not bother to learn our foods and our culture. You may think we are weak because we are vegetarians, but we are strong.*

I love picturing Gandhi finding Salt in a vegetarian restaurant, like readers finding *Satya* in NYC—that chance encounter with words on the page, which sparks a connection, nurtures compassion, and spurs action. I see this moment of discovering an animal rights essay in a vegetarian restaurant as the beginning of a series of experiments with reading that led to experiments with truth, that led to Thatha's own experiments, which led to you, and to me.

---

I research Gandhi's role as editor and the importance of his newspapers as activism. I learn that satyagraha is coined in a contest in his newspaper *Indian Opinion* in South Africa. Gandhi is searching for a name that can unite seemingly disparate goals, all of which should be attained without violence. *Passive resistance* sounds weak, so Gandhi announces a contest. Gandhi's nephew first suggests sadagraha—*firmness in cause*. Gandhi tweaked it to satyagraha—*firmness in truth*.

In South Africa, Gandhi pours his personal savings from his law practice in Johannesburg into the paper he was publishing in Durban and is troubled by the finances of keeping independent media afloat. Gandhi befriends Henry Polak, a local journalist in Johannesburg who frequents the same vegetarian restaurant. The two bond over their mutual admiration of Leo Tolstoy. Polak gives Gandhi a copy of John Ruskin's *Unto This Last*, which Gandhi recounts in his autobiography as having read on an overnight train ride. Ruskin's essays on labor and economy immediately inspire Gandhi to move his press from Durban to a rural self-sustaining ashram called Phoenix Settlement, about twenty kilometers north.

I am noticing another contour—these stories of vegetarian restaurants as sites of encounter that spark change or community, that thing after our own hearts, from Gandhi finding Salt in London and befriending Polak in South Africa to Beth and Martin plotting an activist magazine and readers finding *Satya*.

I think about the days the printed *Satya* magazines arrived, when Beth would run down the stairs, shouting "Ed's here!" Ed, the man with a van, picked up the papers from the printers and stopped by the *Satya* Brooklyn office first to unload the copies for mail subscribers. He escorted the rest of the papers to the other vegetarian/health food stores and community activist spaces in the City. The *Satya* crew would run out of the office to greet Ed, who was often double-parked in the street outside with the back of the van open. Mo usually climbs in the van to pass stacks back to us, and we set up a relay system, back to the office, where Kym's dog Jack waits for us. Each stack is a hundred magazines bound together. We put them in piles in the center of the office. Once our relay is complete, we congregate in a circle around our newly created city of magazine towers, rush to open a stack, and flip through each page, proud of what we made together. At this moment, *Satya* is transformed into something larger than our individual labors and will travel from New York City to six continents.

Many of the advertisers are our beloved NYC community. MooShoes, the all-vegan shoe store, which at the time is located at the site of a former butcher, usually graces our back cover.

My NYC map is marked by meals and conversation of this era, of the places mentioned in *Satya* and where *Satya* can be found.

I know New York City is always changing and that's in part what makes it what it is. Your New York City in the 1970s was befriending hippies. It was taking Mom to the Pan Am Building to watch planes take off at JFK. It was the two of you hanging out with your new Gujarati friends near Lefrak City on Saturday nights. Your friends rented out a high school to screen Hindi movies. Afterward, there was a falafel shop where you would dine. I have looked for this spot, but suspect it is gone.

So many of my favorite spots are also now gone, but I hold my vegan map of the city at that time. The East Village is marked by the dragon bowl special that comes with a slice of Southern cornbread with miso tahini dressing, and a cup of miso soup, from Angelica's Kitchen.

Atlas Café was where I would go late at night after whatever dinner, show, movie, or evening wanderings, to get a slice of Peanut Butter Bomb or Brownie Cheesecake (two desserts in one) baked by Daniel Konya of Vegan Treats, who delivered these specialties each week on Tuesdays.

The vegan buffalo wings and milkshakes at Foodswings in Williamsburg were the subject of my first restaurant review at *Satya*. Sacred Chow in the West Village was a quiet spot where I suggested to meet to conduct in-person interviews.

Every now and then, Kym, Mo, and I would oversee the Brooklyn distribution of *Satya*. Kym would drive, and Mo and I would pop out and deliver the magazines. We made it an eating crawl, hitting up all the best spots, and bringing back lunch for the office. I loved visiting the Veggie Castle on Church Street in Flatbush. It was the site of a former White Castle converted to a Caribbean joint, with collards, jerk tofu, plantains, and juices.

These sites continue to be transformed. A few of the old favorites still exist, and new ones pop up every day.

Nowadays, Beth and I discuss the East Village of the '90s and the '00s and our frequented veg restaurants that are now gone. When we unbury a covered stream, we call it daylighting. I want to daylight these streams of belonging and inclusion that coursed through our city. To daylight our dreams, secrets, joys, and sorrows shared over bountiful feasts, until the tables were topped with chairs and the Open signs were flipped over.

---

In the New York Public Library, I find a collection of **Gandhi's** letters to and from an ailing Tolstoy from 1910, months before Tolstoy's death. The men share their writings and mutual admiration. Gandhi sends Tolstoy a copy of *Hind Swaraj* and his newspaper *Indian Opinion*.

Tolstoy replies:

I have received your Journal *Indian Opinion* and I am happy to know all that is written on non-resistance. I wish to communicate to you the thoughts which are aroused in me by the reading of those articles. The more I live—and specially now that I am approaching death, the more I feel inclined to express to others the feelings which so strongly move my being, and which, according to my opinion, are of great importance. That is, what one calls non-resistance, is in reality nothing else but the discipline of love undeformed by false interpretation.

Gandhi was able to tell Tolstoy about his plans with his collaborator, Hermann Kallenbach, to start an ashram in South Africa. "No writing has so deeply touched Mr Kallenbach as yours; and, as a spur to further effort in living up to the ideals held before the world by you, he has taken the liberty, after consultation with me, of naming his farm after you." This farm would have oranges, plums, and apricots, and be a home for satyagrahis of all walks and religions.

I think of Tolstoy Farm and Phoenix Settlement as the precursor experiments to Sabarmati and Wardha, and of course Kallakurichi.

On choosing to call his first settlement Phoenix, Gandhi writes:

"Phoenix" is a very good word which has come to us without any effort on our part. [. . .] Its significance, as the legend goes, is that the bird Phoenix comes back to life again and again from its own ashes, i.e., it never dies. The name Phoenix [. . .] serves the purpose quite well for we believe that the aims of Phoenix will not vanish even when we are turned to dust.

I am thinking of all these ideas rising from the ash. I wish I could speak to you about these literary lineages I am now seeing. How

Tolstoy and Ruskin and Salt influenced Gandhi, who in turn inspired Thatha to start the Kallakurichi Ashram, and Beth and Martin to start an NYC magazine.

---

When satyagrahis became imprisoned in South Africa for refusing to register and carry identity certificates, as was required by new discriminatory laws, *Indian Opinion* encouraged readers to memorize sections of the paper to relay passages to prisoners when they went to visit them. I think of the title of the poetry collection you give me: *Memory Work and Appreciation*.

While a jailed satyagrahi himself, Gandhi reads Henry David Thoreau's "Civil Disobedience." Excerpts of Thoreau would be sprinkled throughout *Indian Opinion*, and his name is invoked to inspire action. In a summary of a community meeting in 1907, Gandhi discusses the question of whether the Indian community will remain united "to end and disobey the law." He adds, "When we shall have resisted the law to the last, we shall be regarded as so many Thoreaus in miniature."

In her book, *Gandhi's Printing Press: Experiments in Slow Reading*, Isabel Hofmeyr contends that Gandhi's publishing experiments aim to slow reading and writing down to the rhythms of the body. Gandhi, back then, is already wary of the speed with which information is transmitted and concerned about how the brain responds to the bombardment of messages. He reprints extracts from Thoreau's 1863 "Life Without Principle" in 1911 editions of *Indian Opinion*. Hofmeyer contends this passage had a profound impact on Gandhi:

> I believe that the mind can be permanently profaned by the habit of attending to trivial things, so that all our thoughts shall be tinged with triviality. Our very intellect shall be macadamized, as it were,—its foundation broken into fragments for the wheels of travel to roll over; and if you would know what will make the most durable pavement,

surpassing rolled stones, spruce blocks, and asphaltum, you have only to look into some of our minds which have been subjected to this treatment so long.

Thoreau, like me, does not want to do asphalt.

Gandhi and Thoreau's fears seem prescient to the information overload we have today.

You leave us without knowing what social media is or the full reach of the internet. Now we can take pictures and get directions from our phones. Every few seconds there is some buzz in our pockets or on our wrists or a ding on our computers alerting of us of a major tragedy, a minor inconvenience, a well-enjoyed meal, a tiny joke, and an unforgiving audience. We keep on clicking rather than thinking. Hofmeyr contends *Indian Opinion* offers a kind of ethical anthology as an antidote to this bombardment of information. "He interspersed news reports with philosophical extracts, and he encouraged readers to contemplate what they read rather than to hurtle forward." She further explains, "In a Gandhian world, such slow reading became one way of pausing industrial speed, and in so doing it created small moments of intellectual independence."

The influences are cross-cultural, and Thoreau looks to the East. Reading the Bhagavad Gita in "The Pond in Winter," a chapter of *Walden*, Thoreau writes, "The pure Walden water is mingled with the sacred water of the Ganges."

Gandhi's well is mixing with those of Thoreau, Tolstoy, Ruskin, Salt, and ancient Indian philosophies. Gandhi pumps this water and sprinkles the teachings in the pages of his newspapers. What is fascinating to me is how Gandhi processes what he reads and turns it into action. I see this in the work at *Satya* too. We employ a similar editorial strategy of juxtaposing seemingly disparate causes to reveal essential connections. We are also processing what we are learning and turning it into action.

Even now I am thinking about all the utopias, from Ravidas's Begumpura, Tolstoy's inspirations, Gandhi's experiments, and beyond. I don't know if Gandhi ever says *utopia*, but rather *Sarvodaya*, meaning *universal uplift*—the same word he uses to translate the title of Ruskin's *Unto This Last* into Gujarati.

In learning about utopias, I'm looking for your Kallakurichi. But I'm also looking for mine. How do I find the matriarchal, non-hierarchical, animal-friendly, make-love-not-war, vegan bonobo ashram?

~~~

In 1969, you arrive in the United States with seventy-five cents in your pocket.

It is only after I meet your sister, Sachi Aunty, and her husband, Rajgopal Uncle, that I learn what opens the doors to America for you: a diamond earring and a change in United States immigration policy.

In 1965, several years into the war in Vietnam, the United States, to fill jobs, lifts its immigration quotas for Asians. Your eldest brother, Sundaresan Uncle, tells you that America needs social workers.

You apply for a visa and are accepted. All your siblings are so proud of their youngest brother when they first hear the news. But when you tell them you need money for the plane ticket, they all start expressing their doubts. "Why do you want to go to America?" they ask. "What will you do there?"

It is Patti who supports you. She takes off her diamond earring, one of the few pieces of jewelry Thatha did not make her give to Gandhi, and sells it. With that, you are off to America.

When you arrive in the United States, all that is left of that diamond earring is seventy-five cents. You land a job at Johns Hopkins Hospital, you start saving from your paychecks, and you send money back to Kallakurichi. Patti's diamond returns to her earlobe.

"Do you know the story of how your parents got married?" Rajgopal Uncle asks me next.

You never told me this story. I ask him to tell me his version—how he is partly responsible for my existence in this world.

You return to Kallakurichi in 1972 on a leave from work. By that time, you are living in the Bronx on Eastburn Avenue and working at the Manhattan Psychiatric Center on Wards Island. You come back to India to find a wife. You ask Rajgopal Uncle if he knows anyone. He is going on a business trip to Bangalore, where he meets with the boss of Mom's sister, Thripura Aunty, who asks around.

"There is this boy from America," they tell Mom.

After college, she teaches math at the high school she attended. Her students give her roses every day. She has multiple marriage offers but declines them. "I'm focusing on teaching," she tells them. What she tells her friends is: "I don't want a Bangalore boy. I want to go somewhere else, maybe Delhi or Bombay." There is a boy from Canada, but his astrology doesn't match up with hers.

After teaching, she gets a job as a telephone operator in 1971. She has the morning shift and has to be there by six a.m. and leaves the house at five. "It was a good experience," she tells me. "They will call and say, 'I want to book a call.' There was the regular call, the express call, then the lightning calls, which will be nine times the charge.

"There was another small Raji and myself," Mom says. "The supervisor says all these Rajis are very smart. So, they put us on the Bombay board. Very busy, always lit up, and you have to be very careful, very diligent because you could disconnect someone and they'll be screaming."

"People would call to find out time," she says. "I didn't speak Kannada. But you had to say the time in Kannada and English. So, I had to know the numbers from one to sixty."

She trains and works for three months.

"Then what happens?" I ask Mom.

Then, this boy from America shows up.

You arrive in a car with your brother-in-law and pull up in front of her parents' home. Her younger cousins are spying on you guys from a distance.

*Which one is the boy?* they wonder. They see a tall, slender man with hazel eyes step out of the car. You take a comb out of your pocket and run it through your long hair. *That must be him.*

Mom's hair is even longer, parted in the middle and pulled back into a long braid that reaches her hips. She is asked to sing something for you. Then you are left alone to talk. You talk for what seems to be a long time to all the nosy bodies surrounding you both, something that isn't customary in these types of match meetings. The younger cousins have never seen this happen before. Neither has Rajgopal Uncle, who later asks you, "Were you interviewing her for a job or for a marriage?"

But it is Mom interviewing you.

"Can you get kothamalli in America?" she asks.

"Yes. The Mexican stores sell cilantro," you tell her.

"What about mallipoo?" She is wearing a bundle of jasmine in her hair.

"Yes," you tell her. Years later, you grow jasmine plants at home and pick the petals to give to both of us to wear behind our ears.

"Are there lots of hippies there?" Mom wonders.

"Yes." There were hippies. You teach some of them yoga.

"What about Hare Krishnas? Is the movement big there?" she asks.

"Yes." You visit their temple once in Brooklyn.

You remind her of all these questions years later. The answers are somehow enough for her to send her family scrambling to get the paperwork for her passport and visa, and hand-deliver wedding invitations to five hundred people in a matter of days. In the meanwhile, she visits Kallakurichi to get the blessings of her future mother-in-law.

It is January 1972, and Mom is turning twenty-three years old. You are twenty-nine. By February 6, you are married, and on February 16, you both are on your way to New York.

"Pray to your favorite god," you tell her. "We are crossing the Atlantic."

Each year without you, I continue to learn new things about you.

Mom tells me how when you first arrive in the United States, people say you can't remain vegetarian here, and that you certainly can't raise children without meat. I tell her people said similar things to Gandhi when he left for London. But Gandhi found Henry Salt, and that reconfirmed vegetarianism for him, and he reclaimed it as part of swaraj.

Mom says Henry Salt's name sounded familiar, and she thinks you had one of his books. She goes to the basement and returns with a folder of poems and a collection of other writings. There is a newspaper photo of you receiving an award for an essay on Lala Lajpat Rai. You write about the early freedom fighters: "These and other brave men squeezed their blood, nerve and sinew for freeing the enslaved India from the yoke of foreign domination. Their lives show that anything great cannot be achieved without some sacrifice."

In Mom's manila folder, there is also a poem, "The Swami Sonnet," by your friend in the Bronx, about "Adi the Yoga man" and his "modest life" on Eastburn Avenue in the 1970s, before you were robbed and moved out of the city.

The book Mom was thinking about wasn't by Henry Salt, but by a man named Henry Casper. In this folder of papers, I find a slim book, with a stapled spine, almost like a DIY zine, called *Verdome*. In 1974, Henry Casper gifts this book to you in New York City: *To Raji and Adi: Wishing you happiness always, H. Carsten Casper. March 19, 1974.*

I then noticed an inscription on the bottom of the page to me: *Sangu, I am sure you will love this book as much as I did when my close friend Henry Casper presented this to me in 1974. This is 'our kind of book.' Enjoy! Dad 2-10-95.*

I don't remember ever receiving this book. Is it something you mean to give to me and never do, or something I disregard too easily in my senior year of high school? When I go off to college, you will always sneak little things from the house into my bags when I come home on visits—a stapler; a bunch of bananas; Parker pens, your favorite. This is your way of showing love. Do you give *Verdome* to me in a similar manner?

*Verdome* is Casper's philosophy on life, which is in part a health treatise and in part an animal rights manifesto. He discusses the value of gomasio seeds, nut butters, and alkaline diets in treating inflammation. You are always interested in natural healing. I think the place you felt most at ease on land in the United States is in health food stores. They represent a certain American dream, a promise to heal an ailing body. But many of those promises come from wisdom in India. They are also places where Ayurveda, yoga, and vegetarianism are welcomed. I think you feel a kinship with those in the West who take interest and value these concepts from India, like Gandhi finds a home in the Vegetarian Society of London.

I remember you walking into the living room while we were watching a music video of Joan Osborne's "One of Us," where her face takes up the whole screen. "I like this song," you say. I think you like her Indian-inspired nose ring. Sometimes I play that song now just to think of you.

In 1995, when you gift me *Verdome*, the band No Doubt releases their video, "Just a Girl," featuring the bare-bellied, platinum-blond lead singer Gwen Stefani, with her powdered face and deep-red, puckered lips. A jewel-studded bindi shines prominently on her forehead. Around that time, henna tattoo is becoming all the rage, along with toe rings and celebrities wearing black kohl under their eyes.

When I am "just a girl" in middle school, a boy sitting in front of me in class one day turns around and asks if we shopped at Grand Union. ShopRite was closer to us, and I told him so.

"But Grand Union has a 'Red Dot Special.'" He waits a second for me to realize what he said, and then begins to smirk.

I remember how Mom puts on her pottu every morning, sometimes drawing it with a bottle of black or red paste to the spot between the eyebrows, the third eye. We also had those adhesive stick-ons. The red felt dots stick on our bathroom mirror, to be peeled off in the morning and reposted at night. I like to rub my hands over those fuzzy third eyes.

I ignore the kid in school and never tell you what he said. The red dot makes me a target of mockery, but it becomes a target of violence for others. Not far from where we live, across the state border in New Jersey, a hate group called Dot Busters starts attacking South Asian immigrants, referring to them as *dot heads*, regardless of religion. *Dots Die* is spray-painted on Indian Americans' properties. Meat is thrown in front of their homes. I learn about this much later.

You receive mixed messages in this country. When you come to New York in the 1970s, hippies embrace Indian culture. Is this how you meet your friend Henry Casper? Like Gandhi finding Henry Salt in London, you find someone who echoes concepts of ahimsa back to you. Casper had a wit and wisdom in his approach to animal rights:

> Now that an attempt is being made to develop a system of communication or "conversation" between man and the dolphin—the most intelligent oceanic creatures, it is hoped that if success is achieved as expected, the dolphin will represent all water-going creatures in a lawsuit against mankind.

The etymology of *Verdome* is not explained, but I like to think of it as a cross between verde (*green*) and verdad (*truth*), like *Satya*. I would have loved to gift you a copy of *Satya* with the inscription: *This is our kind of magazine.*

I saw in you great benevolence toward all creatures, which sometimes manifested in a sadness about our own species. We never talk about how you cope or don't cope with terrible cruelties to all beings. How you leave everything you know in India for a life in the United States. How you make friends with everyone you encounter, striking up conversations with strangers.

How sometimes strangers are mean, and tell you to go back to your country, or speak to you as if you don't know English—your fourth language of fluency, the one in which you recite Shakespeare and Tennyson by heart.

You protect me from knowing your sorrows.

I have spent all the years since you died trying to get to the source of them, filling pages, asking questions, and searching for answers about how we create sanctuaries in ourselves, in our families, in our cities and wildernesses. How do we create sanctuary for our ethics, when we are confronted by so much violence that is normalized?

*This* is *our kind of book.*

# LOVE IS WORK MADE VISIBLE

**W**HEN I STUDY SOIL MECHANICS IN GRADUATE SCHOOL, I learn about quick conditions; when an upward gradient of water pressure exceeds the submerged weight of the soil, the liquid earth can envelop you.

I see a malaise envelop you from time to time. It keeps you up late at night. You are distant in your private thoughts. But every now and then you talk about failure.

"What do I have to show for my life?" you lament when I am in high school.

You have us, but that is not enough. You have devoted your life to social work. There is still some deep dissatisfaction that is unnamed. You had big dreams—to write poetry or be a journalist or travel in foreign service. Was that your junoon? But you stopped writing. Why?

In soil mechanics, quicksand is called a failed condition. The trick is not to panic. To lie on your back and float.

You were born in proximity to water at that bend in the Gomukhi River. In Kallakurichi, Thatha dug a well that pumps into a big basin where you learned to swim. In the suburbs of New York, you go to the gym at Jack LaLanne. You carry that small pink duffel bag with you to work, and you go there after to swim and sweat in the sauna.

Water always brings you solace. If there is a beach, you wade in, dip your cupped hands in the water and then raise them above your head, like an offering to the sun, to your gods, and then pour it over you.

If I have one survival skill, it is knowing how to float on my back in water, which I learn from you.

You have other knowledges that I never learn. "Look at your dad and all his siblings," Mom says to me. "Their hands are never idle." She is thinking of Savi Aunty knitting blankets. Of Jaya Aunty sewing blouses. Of you finely packing items. Of tending to plants.

There is this gospel of work you all learn in Kallakurichi.

"Work is Worship," Savi Aunty says. "If the eye sees, the hand must fix."

"Work is God. Rest is Change in Work," Jaya Aunty says. "If the eye sees, the hand must do the work."

What happens when what the eye sees is too much for the hands to fix? Is this the source of the malaise, that autumn in the heart/mind that envelops you?

Perhaps it is what envelops me, too—these questions of how to live when there is a willful denial or forgetting. A normalization of suffering. Living sometimes feels like being complicit with it.

There is a phrase that I learn much later that I don't have then: *moral injury*. It's used at first for soldiers, but can expand to other areas, any place where our work is not aligned with our ethics.

Before joining *Satya*, I worked on a Superfund site cleanup in New Jersey. The fierce winds pierced the skin of my cheeks. I hovered behind a work truck, while I waited for a driller to place a core sample on the field table for me to log. I shook a packet of granulated activated carbon, and the portable heat warmed my pockets, where I kept my hands until they were called to document the history of this place.

It started with lanterns. A guy named Dr. Carl Auer von Welsbach figured out that dipping gas mantles into a thorium mixture made them burn brighter. In the 1890s, the Welsbach Factory in Gloucester City, New Jersey, started using this radioactive material in their lanterns. The mantles were handmade, hand dipped, and hand inspected. I don't know the fate of all those hands, but the thorium they were exposed to is still around today. The manufacturing waste was dumped as fill and scattered in parts of Gloucester City and neighboring Camden.

Over a hundred years later, we were trying to undo the damage of Welsbach's discovery. We sampled for thorium in people's basements. We drilled borings in their backyards, in ball fields and parks, trying to figure out the extent of the contamination.

The boreholes were almost all the same. Between the topsoil and the native sands and clays was a layer of fill. An abrupt line marked *before* and *after*. My eyes took note of the various particles among the ash. Black, shiny micaceous flakes. Chunks of creosote. Fragments of bricks. Artifacts of the industries that inhabited the town. My hands, shielded only by blue latex gloves, took inventory.

This was a Superfund project, part of the federal government's program to clean up hazardous waste sites. But calling it a cleanup isn't quite right. Radioactive materials aren't something you can clean. They exponentially decay. We measure that time in half-lives. For thorium, it is on the order of billions of years. We can't clean this up.

All we can do is dig up sections of New Jersey and send them to Utah.

My colleagues numbed themselves after long days. I wanted to do something else, something more, though not sure what. I know I want to devote more of my life to animals.

I didn't want to live numb and small.

When I am at UC Berkeley, a guest lecturer visits us one day to teach us about failure. He makes a business out of it. As a young engineer, he oversaw the drilling of three boreholes into competent bedrock in the San Francisco Bay Area. He assumed the whole site is underlain by this rock and designs a structure above it. The structure collapses. In his guilt, he becomes obsessed with understanding why. He shows his original cross section with the three holes in rock. His mistake is assuming that everything in between them was the same. He overlays the reality—that the bedrock is part of the Franciscan Complex, which is not homogenous but rather a block-in-matrix mélange of particles, some of which act more like soil than bedrock.

The takeaway: don't assume the information you have is the whole truth. Oftentimes our societal pillars are founded on unstable ground. Or maybe, it is that failure is only failure if you don't learn from it. Sometimes in writing this story, I wonder if I'm drilling holes in all the wrong places. I'll never have the whole story. But also, whatever I do find has its own interesting story, and I make unexpected connections.

When the chimpanzees make errors in their sign or use a word inappropriately, their "mistake" often demonstrates far greater awareness than vocabulary. Scientists put objects like bibs and brushes into a box for Washoe to sign, which she does successfully. But for CAT, DOG, COW, BIRD, and CAR, they use small replicas. If she sees these in the wild or in photographs, she makes the signs correctly, but when a matchbox or figurine miniature of them is put in a box, she signs BABY. Maybe this is what I said in Tamil about the elephant statue at the neighbor's house in Bangalore when I am young. Sometimes failure is just a matter of perspective.

# EYES ARE SYMPATHETIC

"YOU ARE GOING TO FALL IN LOVE WITH PIT BULLS," KYM tells me, one evening.

At the *Satya* office in Brooklyn, which is our publisher Beth's house, someone always knows someone trying to find a home for an animal. One December day, Cat sends me an email with the subject: *Hmmm . . . The Perfect Doggie for You?*

A friend of a friend of a friend finds this beautiful white-and-brown spotted dog, chained up to a fence and left out crying on one of the coldest nights of the year. Because of her spots, he calls her Moo Cow.

We go to meet her at her foster's home. We bend down to greet her, and she licks my face and runs to another room to bring us a toy. Kym is right. Wan and I fall hard for Mookie.

Soon, I take her to Prospect Park in the mornings before work. I walk to the *Satya* office and walk home at lunch to take her out. She runs with Wan as he trains for marathons. She comes with me to outdoor-activist events. And she finds her way to the pages of *Satya*, inspiring interviews with other pit bull advocates.

I never tell anyone about the clairvoyant, who visits us right after your wake. She is a friend of your friend Yvonne.

She comes with a message from you: "Tell Manu, I'm happy about the car." Random and specific—coded amends with your son. At the time the two of you were sharing a car. Does it mean that you are happy Manu can take it now?

I am skeptical. Why don't you contact us directly instead of going through this stranger? Yvonne says it is because we are blocked, and you find the next available person.

"Don't worry about me," you tell the clairvoyant who repeats it to us. "I will return as a member of the family."

Then, everyone looks at me as that vessel for your return.

What kind of burden is that—to birth your dead father?

In the years since, my body has not been a vessel for any body. Mookie, who you would have loved, is born probably the year after you left, and comes into our lives two years after that. Early on we experiment with a gentle leader harness that leaves a mark on her nose, similar to the one your sleep apnea machine gives you, and I am suddenly reminded of that clairvoyant.

I want moksha for you, and Mookie is her own person, but I still think of you often with Mookie. When one of her eyes gets infected, I would administer drops in it, and also to the other, because according to you, "Eyes are sympathetic." What one eye feels, the other must too.

# AN ATTITUDE OF GRATITUDE

Jaya Aunty sees I am romba sensitive, like you. She arranges for us to go on a temple pilgrimage to Maharashtra. We take an overnight train from Bangalore to Pune. I read *Dreaming in Hindi*, a memoir by Katherine Russell Rich. In her study of the language, Rich notes there is no word for *privacy* in Hindi. When we arrive the next morning, Mom, Jaya Aunty, and Tripura Aunty are all asking about my lack of bowel movements since traveling.

Jaya Aunty performs Reiki and encourages an attitude of gratitude. "You must love and appreciate your body," she says.

Jaya Aunty is a good role model for this. She divides her day into three parts—eight hours for self, eight hours for home, eight hours in service of others. She takes care of her body, massages her legs, and keeps fit. She constructs a hula hoop made of bamboo and smiles, shakes her hips, and hold her hands behind her head. In her seventies, I see her climb onto the roof of her home to collect fruit from a tree. I am so nervous watching her. I worry she'll get hurt trying to impress us, but she has the confidence and tenacity of a girl raised in Kallakurichi.

Later, I read a booklet created by a friend of hers compiling all of Jaya Aunty's wisdom. She refers to Jaya Aunty as the "Nectar of Grace." Here are her sweet offerings, which Mom translates for me from Tamil:

If you cannot do it today, take rest.
Whatever you do, do it jollily.

You all give too much importance to society. You are always worried about what society will think. That is your problem. Because of this, you are making your life complicated.

The anger, hatred, and sorrow, we collect is dirt.
We clean the house with soap and water to get rid of dirt.
In the same way, you have to clean your mind. Clean it right away.
Whichever god you like—keep chanting their name.

You always live in the past.
Forget the past.
What happened yesterday, went with yesterday.
Nobody is responsible for your happiness or unhappiness.
It's your own creation.
If you carry bitterness, it will only affect you.

Atma can talk to Atma. Soul can talk to another Soul.
If someone is hard for you to face, talk to them in your heart. Tell them, "Sorry. I Love You. Thank You." You will change. So will they.

Mom loves this one. This is how Jaya Aunty blesses her:

May you live long with the strength of an elephant and an iron body.

# AN ATTITUDE OF GRATITUDE

> May you live with wisdom, and prosperity.
> Sandhosama Irunga—*Always be jolly jolly.*

Jaya Aunty thinks I am too much in my head. "Close your Mind," she says. "Open your Mouth."

We are now traveling by car from Pune to Shirdi. On the way I see a sign for a wild animal sanctuary. I ask to stop—this is my kind of temple. Among grassy plains, there are birds, mongoose, squirrels, deer. Goats forage across the street. One looks like Mookie. G. S. Uncle, who is quick with his puns, calls it a Mild Animal Sanctuary.

Sugarcane fields dot the landscape in this region, and along the road there are stands for sugarcane juice. "Want to try?" Mom asks. In Bangalore, machine labor crushes the cane. We don't realize, here they rely on animals.

We meet a bull named Raja who is tied to an apparatus. He has rope through his nose and chains on his neck. He is only able to walk in a circle, which powers the device that crushes the cane to collect juice.

Raja wears the heavy chains even when he is resting, and is confined to the small radius around the crushing apparatus.

A family operates this stand. A little boy with a stick pokes Raja. *Stop!* I suddenly scream.

I ask if they can remove the chains. G. S. Uncle, trying not to offend this poor family making a meager living off of sugarcane juice, makes the request. I am a tourist, he explains, wanting a picture. For a few minutes, Raja is relieved from the weight on his body.

Manu and I wonder what you would have done in this situation. How do you balance expressing concern for Raja and extending compassion for this family? This is one sugarcane juice stand in a region filled with them. Would this have haunted you the rest of the journey?

What happens when the eyes see, and the hands cannot fix the sequence of failures we encounter daily?

When we arrive at Shirdi Temple, the stray dogs approach Jaya Aunty. She does Reiki on them and offers them temple prasadam, but they are too sick to eat. She sees you in me and wants to teach me a lesson.

"Adi cares about dogs and people, not fancy clothes, or society. If there is a dirty sick dog, Adi will go and carry him and not care that his clothes are dirty." Jaya Aunty sits with a dog on her lap, dirtying her saree, as she tells me this. "Dog is important. Saree is not. But society cares about sarees not dogs. To get society to care about the dogs and people, Adi learned to wear nice clothes and get the degrees. If they respect him, they will respect what he respects. Do you understand what I am saying?"

The way Jaya Aunty says *society*, it is as if it is outer space. "Kallakurichi was Jolly Jolly," she tells me. "Did not yet know suffering."

You are raised on the principle of Jeevakarunyam—*compassion for all living beings*.

"Society is hard," Jaya Aunty says, acknowledging that society doesn't operate on the same principles as Kallakurichi. She turns to prayer, to recite shlokas a thousand times, to calm her mind.

When the eye sees, and the hand must do the work, what is the work that I have chosen to do?

# WORK

**F**IRST, IT IS TEACHING MY HANDS TO SIGN.

I think back again to my junior year in college, when I applied to the Chimpanzee and Human Communication Institute to spend my summer in Ellensburg, Washington, with the signing chimpanzees. Do you remember, I am also accepted into a soil and water program at the space center in Huntsville, Alabama? I think I can find other opportunities to pursue soil and water later in life, if I want to, but I don't know if I would ever have the chance to meet chimpanzees. I also am curious about going to a place that is reckoning with the ethics and the aftermath of research experiments with chimps.

You don't understand at first. You want me to choose the space program. The space program is paid. The chimp program you have to pay to be an apprentice.

Why do I want to pay money to clean up chimp shit? you ask.

I think you, of all people, should understand.

We fight over this. That is around the time of my first colitis flare-up, another inheritance.

Is my emotional anxiety about disappointing you the reason for the flare at that time?

I choose the chimpanzees but get a summer job at the Cooper Union Research Foundation the month before this apprenticeship to save money and pay for the program.

CHCI is an attempt at making amends with a past mistake. Roger Fouts admits it was wrong for scientists to snatch chimpanzees from the wild, breed them in captivity, and separate them from their offspring in the interest of science. Since they are raised as human children, they have no place they can go back to and survive on their own.

Fouts has his own Ashoka-like transformation, and CHCI is an attempt to give these former language research chimps as much dignity, agency, and joy that a life of captivity can provide. The institute has multiple enclosures. The large outdoor space is about seventy feet high to give the chimps the opportunity to climb like they would in the forests. When the weather is too cold, they have the option of going to the indoor playrooms, which have tires, swings, and toys. In the evening, they go into "night cages," where they eat dinner and make nests with blankets.

A day at CHCI is about setting up different rooms with food, treats, forages, games, puzzles, and opening the doors to let the chimps in, and then locking them in there while staff and volunteers clean the other enclosures. As a summer apprentice, we rotate between these various chores.

I hose the night cages, douse them with chlorine, and brush feces down the floor drain. I make sure each enclosure is finely scrubbed down, and the drains are cleared of clumps of chimp hair.

The director of the lab, Mary Lee Jensvold, orients us volunteers. "We can choose to walk away. The chimpanzees do not have that choice. They cannot go back to the wild."

There is an inherent cruelty in captivity. "They should not have to sit near their own shit," Jensvold tells us. She makes the unglamorous assignment of cleaning chimp enclosures feel noble.

CHCI is not a zoo. It is an anti-zoo. Roger Fouts tells us that all zoos should be working to put themselves out of business. As an educational institution, CHCI conducts these once-a-month Chimposiums. The public listens to a lecture about chimpanzees, their plight in laboratories and in the wild. The Chimposium concludes with a visit to the indoor gallery that looks into the chimp enclosures through glass.

Chimps are territorial creatures and exhibit threat behaviors. Three or more threat behaviors is called a display. Dar makes himself as big as possible, hair sticking up, in a bipedal swagger, teeth showing as he punches the shatterproof glass. Visitors are instructed to cower low and show a pronated wrist in submission, until the chimps calm down and realize they are not a threat.

Some of the chimps are curious about the visitors, while others move away and do their own thing. Someone once asked if CHCI should use one-way glass so as not to upset the chimps with the presence of humans. CHCI believes these are natural chimp behaviors and they deserve the right to know who is watching them and react appropriately.

I love it when it is my turn to be on berm patrol at CHCI. Whenever the chimpanzees are in the outdoor enclosure, a person is stationed on the berm, an elevated walkway embankment that offers a view of the chimps and the surrounding settings. Up there, I watch the chimps, and they can interact with me, if they choose.

I write my observations in a field logbook, feeling like the Jane Goodall of Ellensburg. As part of our apprenticeship, we learn chimpanzee behavioral taxonomy and are given a blue booklet that describes and codes various chimpanzee behaviors and facial expressions. AA is *Arm Around*, when one chimp puts their arm around another. PLF is *Play Face*, when the upper teeth are covered, and the lower lips expose only the bottom teeth, and is accompanied by a nodding head. This should not be confused with grinning, where all teeth are shown, signaling fear. We note if the chimps sign to themselves (*Private Sign*) or to others. Out on the berm, the female chimpanzees come to check me out. Washoe signs SHOES and GIMME. Moja likes to cover her face and sign

PEEKABOO or cross her hands over her chest and sign HUG/LOVE. Tatu rubs her index finger on her chin, soliciting CEREAL. Tatu loves games, puzzles, and flipping through magazines and pointing out everything BLACK. But every now and then, I find her sitting and rocking back and forth, a captive behavior that started decades earlier when she was separated from her mother. "When rocking, chimpanzees periodically stare straight ahead as though fixated upon an invisible point in space. They are, however, quite aware of everything going on within their field of vision," the blue behavioral taxonomy booklet notes. "Tatu can often be seen low arousal rocking before regurgitating."

Again, I wonder what to do when the eyes see, and the hands try but cannot fix?

# THINK-TROUBLE

After I start working in San Francisco, I look for other opportunities to volunteer with primates. *Work is God. Rest is Change in Work.* I know Koko the Gorilla lives not too far away, and I send an inquiry about volunteer opportunities at the Gorilla Foundation.

I am twenty-three years old, and I follow the instructions I am given for my interview: *Dress casually. Wear comfortable boots. When you get to the plateau near the top of the hill, stop at the gas station, a motorcyclist hangout. Your cellphone won't work. Call using the payphone and await further instructions.* I wonder if they are going to blindfold me from there, to protect the secrecy of the location. I drive further up the road, as instructed, until I reach the gated driveway with the sign: Beware of Dogs. Two gorillas, Koko and a male silverback named Ndume, live there.

I never have direct contact with the gorillas. Instead, I am given a data entry assignment. I work remotely, inputting the old sign language logs of another gorilla named Michael, who has recently passed away. I input phrase acquisition data and note how many signed words Michael strings together, and if the words are spontaneous and unprompted or a response to a question.

What I know at the time is that Michael was born in Cameroon and orphaned, like my small smalls, due to the bushmeat trade. I am told Michael is an artist and a sensitive soul. He paints and names his art works. *Apple Chase* is a painting with white and black strokes that resemble the black-and-white dog named Apple with whom he would play chase.

Both Koko and Michael use the sign FAKE to express doubt and disbelief. I love that they are also seeking truth.

Michael remembers the death of his mother and signs about what happened. He tells it to his caregivers, who film and transcribe it:

SQUASH
SQUASH MEAT GORILLA
MOUTH TOOTH
CRY
SHARP-NOISE LOUD.
BAD THINK-TROUBLE LOOK-FACE.
CUT/NECK
LIP GIRL HOLE

What the eyes see, the hands tell.

# SOILED HANDS

**I**T IS THEIR HANDS THAT TELL ME WHAT HAPPENED.
In July of 2005, I travel to Rwanda with my friend Rachel from *Satya*. I meet up with Rachel in Nairobi, where she is living at the time, and we take an overnight bus to Kampala, Uganda. We rest for an afternoon, eat ground nuts, greens, and sweet potatoes before starting another twelve-hour ride to Kigali on the Jaguar Bus Line. Nollywood—Nigerian Bollywood—videos play on the bus, while we look at green fields of sugarcane. At a rest stop along the way, we purchase samosas and pineapple chunks on skewers—better vegan fare than we can find at any rest stop in the United States.

At the Rwandan border, we cross over on foot, carrying our pineapple chunks in our hands, and enter the Land of a Thousand Hills. There is an old Rwandan saying, that God sleeps in Rwanda because it is the most beautiful place in the world. Or perhaps God must have been asleep during the 1994 genocide, when nearly a million people were slaughtered in a mere hundred days.

Unlike my trip to Cameroon, I don't arrive in Rwanda with a big fat white binder. Instead, I carry a small, fat, black Moleskine journal that fits into my pocket. "Is that your Bible?" Rwandans ask, perhaps because it is so thick.

I had recently left my engineering job for *Satya*. I was figuring out my role and identity as engineer, journalist, activist, global citizen.

After my trip to Cameroon, I read more about war and forests. One day, I am at a local activist bookstore collective in NYC, Bluestockings, and one of the collective members suggests another book on conservation in areas of conflict. I arrive in Rwanda with a shrunken map, a list of a few contacts, the names of forests, and questions in my little black journal. My questions are big and do not have easy answers. What does a country look like eleven years after a genocide? How does a nation heal? If a bible is a book of authority, what I carry is a book of questions.

Rwanda is home to chimpanzees, thirteen species of monkeys, and the famous mountain gorillas in the mist. To see them in their natural habitat—on their terms—is partly why I am there. I spend the previous several years seeking opportunities to visit and volunteer at primate sanctuaries as an escape from my steady environmental engineering day jobs.

It has been seven years since I first met Washoe, and I am still chasing apes. I can't quite explain it. Maybe it is because Moja crosses her arms over her chest and signs HUG/LOVE. Maybe it is because I play chase with Kiki Jackson in Cameroon, and when we get tired he presents his arm for grooming, and I pretend to eat the imaginary bugs I find. Or because I like watching Nama's fingers lace and unlace the shoestrings of my hiking boots. Maybe it is because I make them lemongrass tea when they are sick and banana leaf burritos when they aren't. Or perhaps it is because I hold and bottle-feed the babies. It is the first time in my life I feel like a mother. I also know I can never replace the ones they lost.

These primates that I've gotten to know in sanctuaries have been rescued from research, entertainment, or the illegal bushmeat trade. Meeting them is a reminder of what has been taken from them, but also a chance to see them start new lives. I want to know it is possible.

To see them play and form new bonds is amazing. For *Satya*, I interview Carole Noon, who has devoted her life to rehabilitating hundreds of former laboratory chimpanzees. She makes reintroducing chimpanzees to each other sound so easy. "You open the door. The chimps walk through the door and do all the rest. I think I can say, every time I've opened a door, they've gladly walked through it. They are so ready to get on with their lives."

For others, it is harder. I am reminded of a chimpanzee I met in a sanctuary in Quebec. She spins her head in figure eights. Used in hepatitis research, she previously lived alone in a cage for eleven years. She has anxiety attacks. Though her laboratory days are long over, she still spins her head in figure eights.

Part of the reason I go to Rwanda is to see free-living apes who do not know such traumas. In my black bible I scribble questions about threats to their habitat. I am interested in what animal protection looks like on the ground.

On my daily commute to my prior engineering job, I read about Rwanda's genocide. I was going through a phase of reading books about human violence. "Don't you read anything fun?" my friends ask. I was, at that time, the definite buzzkill at parties.

As an engineer, I studied catastrophes—earthquakes, landslides.

Rwanda has eight million people—the population of New York City. What did it mean to have that population decimated? What was the half-life for the traumas of a genocide? Unlike my work on that Superfund site in New Jersey, the damage can't be isolated and removed. There is no Utah to send it to.

---

One of the first things we do in Rwanda is visit a genocide memorial. We take a minibus to Nyamata, which plays the song you like: Joan Osborne's "One of Us." As we get off the bus, we meet a Ugandan civil engineer who is rehabilitating the Ntarama Church Memorial.

We walk a few kilometers through banana groves with their purple blossoms and pass goats and long-horned cows. The red laterite roads kick up dirt onto our calves and dust our eyelashes strawberry blond.

The church is a construction site. A Rwandan survivor named Pacifique opens the gate for us. The first thing I notice about this slender man is the scar on top of his head. Pacifique points to his head and says, "They slashed."

During the genocide, people gather at this church for safety, but find no haven among these wooden pews or under this vaulted ceiling. Pacifique was left for dead. He survives, but the rest of his family—"enfants, ma mère, mon père"—are killed here along with five thousand others. This church in Ntarama is nothing like the memorials of sculpture and stone that I have seen in the United States. It is a memorial of wood and bone.

When we enter the church, we walk along the aisles between wooden benches. The personal items of the deceased are also scattered about the floor: little girls' shoes, scarves. A rosary on top of a mattress. God sleeps in Rwanda.

The bones are sorted and stacked. Piles of femurs fill one corner. Skulls with vacant eye sockets gaze at us from nearby shelves. Skulls—too many to see—are also heaped in a bag against the wall. Baby skulls slashed in half. Baby skulls slashed into multiple pieces.

Rachel and I hold each other and weep. The Gayatri Mantra, the only Hindu prayer I know, the one I learn from you, plays on autoloop in my head; my instinct is to pray, but my prayers—like those of who sought refuge in this church—are inadequate.

The room adjacent to the church is a school. A child's notebook lays open on a small desk. Pages holding the cursive of little hands writing history flap in the wind.

I look at the guest log of the church. The news correspondent Katie Couric was recently there. Later, I begin to think more about memorials and who they are for. Years later, I meet a young man from Côte d'Ivoire, who was forced to fight in his nation's civil war. We talk about Rwanda and this memorial.

He asks why the bodies are kept on display.

"I guess as proof that it really happened," I say.

"No. They should bury those bodies," he insists. "They won't forget."

What the eyes see, the hands must bury.

~~~

A few days later, we travel to the northwestern corner of the country to Ruhengeri via minibus. It is a beautiful ascent, and we are surrounded by the Virungas, the volcanoes whose names I have seen in so many books: Sabyinyo, Bisoke, Karisimbi. They resemble the mountain gorillas who call them home. We come to see the mountain gorillas in the gorilla mountains, straddling the borders of Rwanda, Uganda, and the Democratic Republic of the Congo.

Rachel and I are complementary vegan traveling buddies. We tag team our language skills. With my basic French and Rachel's basic Swahili, we find greens, beans, grains, and fruit. We'd ask for l'eau chaude or maji moto—*boiled hot water*—that we pour into our own bottles, avoiding plastic waste. Rachel is perceived as muzungu and I as muhindu, but people always ask if we are sisters. "You look similar." After a while, I start saying yes. That is how I feel.

It is a rural town, with many bicyclists. Schoolchildren follow us asking: *What is your name? What is your job?* In Rwanda, it isn't me who has the difficult name to pronounce, but Rachel.

"Recho?"

"No. Rachel."

"Oh, *Vittol!*"

The night before our much-awaited trek to see the gorillas, I get altitude sickness. I can't sleep and am nauseous. I puke beans, while Rachel holds my hair back. We go to bed and wake up bright and early. I am not feeling fully recovered but can't miss this much-awaited gorilla dream.

When we arrive at the park, they divide us up into groups to decide which gorilla family to visit. Susa is the biggest group, and it is a strenuous hike to find them, the way I imagine trekking in the Virungas. Mount Sabyinyo, which looks like a gorilla chin, also fills up quickly. We are assigned Amahoro, which means *peace* in Kinyarwanda. It is a group of fourteen gorillas. We start at the base of Bisoke, walking through pyrethrum plantations until we reach a forest. I expect to hike for a few hours before reaching the gorillas, but they appear in minutes.

At first, we see fuzzies in the trees and get so excited, and then are surprised by the silverback named Ubumwe (*unity*), who walks right by us, showing his big gorilla butt. We watch them eat and chew on bamboo. We see little ones stumbling over themselves. It is sunny, and when the gorillas get too hot, they move to the shade. We see the most recent young addition to the group, a female named Rwanda.

For me, being in their company is also like a child getting a thing after her own heart.

In Kigali, we visit the newly created Kigali Institute of Science and Technology, KIST for short. A sign reads: "Is any African University really contributing to the sustainable development of their countries?"

Although I had recently quit engineering to join *Satya*, I haven't abandoned it. I am still searching for meaning, purpose, and possibility. I want to know how science and technology are being applied in the rebuilding of a country.

At KIST, we watch a video about biogas being used to fuel prisons. Rwanda has a huge prison population awaiting trial. The year we arrive, Rwanda starts gacaca, *justice on the grass*, a community-based truth and reconciliation process to repatriate prisoners back into society.

In many towns we can notice gacaca from afar. What we see are gatherings of people sitting under rainbow umbrellas outside.

We continue our travels to Butare, where we stay with Felix, a professor at the National University of Rwanda who is friends with a professor friend of mine at Cooper Union, who connects us. All of the faculty at NUR were killed during the genocide. Felix is from Ghana, and he has lived and studied in Bulgaria and Brazil and came to fill the need for professors in Rwanda.

He brings us to the university and introduces us to several departments. He finds it interesting that I was an engineer and am now a writer, and that Rachel had studied nutrition and politics in college. "In America, you have such unique combinations," he remarks. Our interests in writing and food come together again when we pick up a student newspaper and learn about a women's cooperative for soybeans in the nearby town of Gikongoro, where they make soy milk, cheese, and ice cream. The prospect of finding vegan ice cream in Rwanda thrilled us.

Each day, we walk back and forth along the main road to the university. In the evening the power goes off, and we make our way back to Felix's house in the dark with flashlights.

We meet with the head of the civil engineering department. My black book fills up with notes on water resources, wastewater discharge, and names of rivers. Butare gets its energy from hydro dams in the neighboring Democratic Republic of the Congo, and the unreliability of that source explains the frequent blackouts.

One evening, we discuss gacaca with Felix and his friend Caroline, who is pregnant at the time. Felix is worried that these stories will harm her baby. Caroline is Kikuyu from Kenya but married to a Rwandan. She thinks it is important for her baby that she attend the gacaca.

The gacaca in Butare isn't held on the grass like in other towns, but in a town hall–type structure. There is a panel of ten judges, seven of whom are women. They wear sashes and sit on the stage. The whole process is conducted in Kinyarwanda, so our limited Swahili and French are of no use here. A man is summoned to take an oath. The panel asks him where he was during the genocide. He has a prepared written statement that at times he reads from verbatim, and at other times he goes off script, making eye contact with the audience and telling what seem to be jokes every now and then.

A very kind young man sitting next to me explains the process. Many people are called to give their testimony, he whispers. It is an opportunity for the judges and the audience to ask questions and gather information. He offers to take notes in Kinyarwanda and records them in my little black bible.

---

Many years later, I return to this little black book and seek out someone to translate these notes. My translator, T, wants to be anonymous. T has legal, ethical, and intellectual concerns about the document: We don't know the distance between what the speaker, S, said and what the notetaker, NT, wrote in my journal. When you transcribe, you lose the stops, T tells me. You lose the voice and tone. We don't know if NT is writing what S said, or summarizing what S said.

T offers a critique of the document. "Not the content, but the system around the document." We talk about reading and confronting other sources. That's how you put the "hygiene" in the material, T says.

T translates what is said and creates dialogue around what is not said.

He notes the indirectness of the language of S. T observes a literacy around power. "Power is something very important. Who has the power has an influence on the speech."

I know T carries his own losses of this time. He was very young when the war and genocide happened, and he is still processing it. He thinks these efforts to understand the past are important, "for the future," he tells me. When we begin our work together, he sets his terms.

"If I do something, you know that I did it, and you acknowledge me in your heart, but you don't put my name under that document," he insists. "In that way, I'm protected. We can be free to do the intellectual work."

I agree, and we commence this work together.

S, called to testimony in the gacaca, begins with "What I know…"

T and I have weekly video calls, going through the document page by page. T notes what is said and what is not said. After each page, we pause for questions and commentary.

T explains nuance. He provides a longer complicated history of the country and its porous borders. How, to understand what happens, you need to be able to track information in space and time and see what was happening both in and outside of the country.

What T can deduce about S is that he was likely a businessman. S tells the gacaca what he did and witnessed during the genocide, and information he has about the deaths of certain individuals.

We learn:

S saves the lives of fifty-two people.

S hides them above the ceiling, feeds them at the hospital, pays large sums for their safety, and begs for his own life.

S tells the gacaca about the ones he couldn't save as well: the ones killed at the match factory in Butare; his brother's grandchildren found in the toilet; and his friend who refused to flee in May, an inauspicious month of floods, disease, and tribulations.

They do not yet have language for violence of this scale.

S says "istembabwoko," combining gustemba (*to erase*) with ubwoko (*category*). Genocide.

"When you go to fetch water from the lake," T says, "you need a container. The container that's the language."

~~~

When Rachel and I leave Butare, we stop by Gikongoro to find the soymilk women we read about in the student newspaper, but it is gacaca day there. It does not appear to be in session, but all shops are closed. Several people are curious about us and try to track down the cooperative owner but have no luck. We buy bananas from a street vendor and wait for the bus to Cyangugu.

Cyangugu is where poop powers a prison. We decide to walk there. At the entrance to the prison yard, I explain that we have heard about the biogas and are interested in learning more, and they let us in. The prisoners are outside working in the fields, wearing their crisp uniforms—pink shirts and shorts. It is a Friday, which is visitor day, and families wait in a long line to bring food and supplies to their relatives. They have a beautiful view of Lake Kivu. At the prison, there's a place prisoners can go to offer their confessions, to participate in gacaca to be reintegrated.

We walk on the same yard as the prisoners and are not separated by any bars or fences. In the field, shrubs are planted to spell out B-I-O-G-A-S. Prosper, the prison director, guides us to the facility that converts toilet waste to cooking fuel. A man props a ground cover open. I peer into the hole. *Is this what you came to see?* the man asks.

I feel like a sponge, taking in what I can, guided by some internal quest that I cannot articulate.

~~~

We continue our travels to Nyungwe Forest, in search of chimpanzees. We meet Clauvelle, who will guide us through the primate-filled forest that day. Rachel asks if our trek will be difficult, and Clauvelle says yes, that "it is difficult, but we are lucky, we are strong."

We start at a high elevation, with fresh mountain air and flycatchers and kingfishers flapping around. The hike is actually a descent. We walk down, down, down to try to catch the chimps in their morning nests before they depart for the day.

When we reach the bottom of the trek, we peer and squint and see one chimp way up in a tree. We can make out her shape and her swagger. There are other nests in trees, but the chimps have vacated them for the day. Monkeys eat figs and drop them on us. One monkey falls way down and lands on a branch and poops on me.

I am not disappointed. To be in Nyungwe is to see the height and expanse of their range, to feel a bit of their freedom in this forest, and I am so happy for them. This is what I came to see.

We start our ascent back through a different route, and still climb up, up, up. We pass villages and farms, pine and cyprus plantations. Then there is a coltan area. Hands mining metals for electronics on other continents. We realize the precarity of this forest freedom, with these other pressures on the land nearby.

―――

On our last evening in Rwanda, we are in a bustling restaurant in Kigali, and I find myself looking across the room to see two Deaf men. I love watching hands tell stories. I have learned the way fingers set up scenes and describe characters. The way mouths and eyes corroborate what the fingers say.

I am unsure of the sign language they are using. In Butare, we meet a boy who uses French Sign Language, which is linguistically very similar to ASL. The men catch my gaze and laugh. A young man comes over to our table and tells us the men were signing how beautiful we were. The sign they use for *beautiful* looks to me like *wonderful* in ASL.

"How do you know what they are saying?" I ask.

"My mother and brother are Deaf," he says.

He sits down with us to finish his meal. The restaurant fills. Music blasts. People laugh. We talk about our travels. The young man is from Rwanda but has spent the last decade in Kenya. I ask him if his mother and brother are still in Kenya. He replies with three words.

"No. Dead. Genocide."

Two pairs of hands signing, two pairs of hands silenced.

# BEARING CAPACITY

What I know about resilience and resistance, I learn from studying soils. When loaded heavily or too quickly, soils can undergo a bearing-capacity failure—they collapse. Even after heavy loads are removed, the memories of previous burdens remain in each grain of clay.

As a geotechnical engineer, I know that soils gain strength over time from withstanding pressures. Sometimes, engineers preload a soil to acclimate it to future loads. Drains are placed to alleviate the pore pressure and decrease the time for the soil to settle. When a future weight is placed on that soil, it will have the memory and strength to carry it.

How do we carry these burdens without collapse? When do we gain strength? Where can we alleviate the pressures?

Can we learn to not put this weight on the future?

# FINDING KALLAKURICHI

After we immerse your ashes in the Ganga, I send a letter back home to friends. "Our journey began in North India," I write, "but this story really begins in South India in a place called Kallakurichi, where my father was born."

Kallakurichi lives in my imagination, my subconscious, my soul. It is a place remembered and forgotten in perhaps equal parts. I believe it holds all the questions and all the answers. I only have to get there.

What I know of Kallakurichi, I learn in bits and pieces of English and Tamil over time.

Savi Aunty tells me that, after leaving Burma, Thatha spends a year trying to figure out where to build this home, this life of service to the community. He looks in all directions, toward mountains and the sea, before settling on Kallakurichi, a name that requires lingual movements in all the cardinal directions.

Uma Aunty's husband, Krishnan Uncle, tells me that the decision to come to Kallakurichi comes to Thatha in a dream. Narayanan inge than Irikku. *Narayanan, it is here only.*

Amirthalingam Uncle tells me Patti's family does not support this radical shift to voluntary poverty and tells Thatha that, if he wants to pursue his ascetic way of life, he should, but to leave Patti and the children out of it. Thatha says there is nothing stopping his wife from leaving, but his children will stay with him. Patti chooses to stay with her children and adapts to this village life.

Jaya Aunty tells me that Thatha and Patti visit Gandhi's ashram in Wardha. Patti comes from a merchant family. Her parents had been trying and praying for years for a child, and when Patti finally arrived, she was so spoiled, rumor has it, that she never had to walk—her uncles would always carry her. But she still learns business smarts and knows how to manage her very large household.

During her stay in Wardha, Patti becomes an administrator of the day-to-day work. Gandhi is so impressed he asks her what she would like as a gift. Patti has two requests: betel leaves from Tanjore, and coffee from the mountainous slopes of Coimbatore. Gandhi makes arrangements to provide this.

Thatha asks Patti where she got the coffee she serves to him, and he is upset. Is this because it is too extravagant for the simple life they are now pursuing? I wonder what is more upsetting to him—enjoying a luxury item in the presence of the Mahatma or that Gandhi lavished praise upon Patti for it?

Gandhi reassures him it is okay, saying it is important to appreciate one's wife.

I wonder if Gandhi heeded his own marital advice. I can only read into the silences to find the unspoken suffering of Kasturba, a wife who seems to have a life of martyrdom imposed upon her. A woman who lost her first child around the time Gandhi's father died. Gandhi writes only, "I may mention that the poor mite that was born to my wife scarcely breathed for more than three or four days." I think of Kasturba's invisible pain.

In his weekly newspaper in 1935, Gandhi writes about his Wardha ashram in Maganwadi being overrun with visitors. "The result is that my abode has become a dharmashala without any private quarters. [. . .] Add to this the feat that we are working without servants. Cooking, washing and cleaning are all done by us. The resources of Maganwadi are therefore truly taxed when visitors come as they do without notice."

Perhaps this is why Gandhi is so grateful to Patti, who is mother to at least eight children by that point and knows how to feed many mouths. She is a visitor who took up the work of the ashram.

Amirthalingam Uncle tells me Thatha models Kallakurichi after Gandhi's Sabarmati and builds the ashram on the banks of a river. "I do not want my house to be walled in on all sides and my windows to be stuffed. I want the cultures of all lands to be blown about my house as freely as possible," Gandhi says in a speech in 1921. Thatha, perhaps, takes this as an instruction manual, and builds your childhood home on these principles, to have no interior partition walls or rooms, just common spaces, with no locks on doors or windows—an experiment in trust.

"You are making me very happy," Jaya Aunty tells me, when I ask her about Kallakurichi on our train ride to Maharashtra together, several years after your death.

"Back to the pavilion," she says. "I lived in my own world. Each of us had our own world."

Kallakurichi means different things to each of you, and to each of us in the next generation.

Jaya Aunty and G. S. Uncle help me sketch the ashram in my journal.

The house has a thatched roof made from palm leaves, constructed by the children and supported by bamboo columns. The outer walls are brick. Your Kallakurichi house is situated in the bend in the river, and surrounded by forest groves—cashew, jackfruit, mango, guava, coconut, and karuvelam maram. Cotton and rice are also planted.

On that Maharashtra trip together, Jaya Aunty spots karuvelam maram along the road. I look up the scientific name, *Prosposis juliflora*. Jaya Aunty says you call it *pis a pis*, with seeds like toor dhal, salty and peppery, thorny with yellow flowers. Native to Africa, their roots are now submerged in the groundwater of Tamil Nadu. You pack bags of them in Kallakurichi to ship throughout the country for erosion control. Can we trace this karuvelam maram back to Kallakurichi?

G. S. Uncle draws the large swing near the entrance of the house. The kitchen sits along the back adjacent to the river. There are built-in shelves for "books of knowledge and utility," Jaya Aunty says. I want to find and read this Kallakurichi library.

When I imagine Kallakurichi, I see groves of mango, guava, and coconut trees, and you, young and nimble, climbing them. You carry a small knife and some chili pepper to slice and spice your treats. Ranjani Aunty visits once and throws stones at the tamarind trees, trying to knock down the fruit. I picture the thirteen of you children washing the thirteen cows with coconut shells, as Savi Aunty describes to me.

In Kallakurichi, revolution begins by learning to spin. You tell me Thatha joins the Tamil Nadu Spinners Association. Gandhi promotes building places to teach cotton spinning and basic hygiene. As children, you "make clouds" by picking cotton, and you all stitch your own blankets, pillows, and mattresses.

Jaya Aunty says Patti sleeps on a cot. You keep an aluminum plate of water and kerosine at the foot of each leg to deter the bugs. You all wake up at four a.m., remove your mosquito nets, fold all your bedding, and bathe in the river.

Jaya Aunty tells me that you each keep three sets of homemade khadi clothes. One to wear, one to wash, and one for emergency or guests.

Spinning is Gandhi's answer to all the evils. It is an act of defiance against imperialism and industrialism. It gives livelihoods to the poor, regardless of caste or religion. The act itself is a kind of meditation: "And those who do spinning as an art know the pleasure they derive when the fingers and the eyes infallibly guide the required count." When the eyes and the hands do the work together.

I hear from G. S. Uncle, full moon night in Kallakurichi is so beautiful. All the frogs come out. Amirth Uncle tells me there are thirteen dogs in Kallakurichi too. Each time one dies, he buries them, says a prayer, and plants a lemon tree. One time a cashew tree. Sachi Aunty warns of the snakes, reptiles, and monkeys. But you all live in harmony with animals. Jeevakarunyam. *Compassion and kindness to all living beings.* This is what I am searching for when I go searching for Kallakurichi.

To get to the Kallakurichi Ashram, Uma Aunty says you have to cross the Gomukhi River. Thatha employed his engineering skills to build a bridge. At the entrance of the property sits a brick building, the peria, *big,* office, where Thatha kept his correspondences, books, and engineering drawings. He has letters to Gandhi and other Freedom Fighters. How do the decisions he has made connect to a larger struggle that is occurring in the country? By finding Kallakurichi, connecting with the land and his projects, I thought I could know him.

---

You all talk of Kallakurichi like a long-lost friendship, with both fondness and regret. But there was something haunting about that place too. Nightmares that wake you up in the middle of the night, decades after you leave. Despite so much reverence for this place, all of Thatha's children disperse to different corners of the country and the world. You all leave and don't come back.

Here's what I have gleaned. You first leave Kallakurichi at the age of thirteen. Like the rest of your siblings, you are homeschooled and ready for the college entrance exam at that age. Patti brings each of you to Hindu Benares University to sit for the admission tests. These exams could have been taken at another location, but Patti chooses Benares because it is located on the holy river Ganga. Every day, Patti takes you to each ghat along the river and dips each of you into the mighty water for a prayer and a bath.

You complete your studies at Loyola University in Madras. Having grown up sheltered on a rural ashram, you have never once stepped into a classroom before going to college. You are still a child. Frail, skinny, and taking classes with students several years older, you don't fit in school. You were picked on by the older kids and are uncomfortable in social settings.

In 1959, when you were only sixteen years old, Thatha dies.

India was post-independence, post-partition, post-Gandhi, trying to figure herself out. You are a boy trying to find your place in his changing country too. Patti says Gandhi came in her youth, and when he left, her youth was gone. Meena Aunty says that Gandhi is the man who makes her mother cry. What do you say?

Thatha's activism puts a social and financial burden on your family that Patti carries long after both Thatha and Gandhi are gone. You, as the youngest child, never know the wealth and comfort your older siblings experienced in their early childhood. Patti compensates for this with affection.

Thatha is always described as more legend than person. I am only told his virtues, but I know there are vices. There is rage, a violent temper that is alluded to; but you and your siblings hesitate from speaking too ill of your late father.

How is it that I know there was violence and nonviolence in the home?

~~~

Do you remember when I was in fifth grade, I played Mark Antony in our class play about Julius Caesar? I wear a satin white fabric, draped like a saree. "Friends, Romans, Countrymen, lend me your ears. I come to bury Caesar, not to praise him. The evil that men do live after them, the good is oft interred with their bones, so let it be with Caesar," I say, on the stage in our auditorium/gymnasium/cafeteria.

When I am ten, I don't fully know what this means. Does it mean that the good things are forgotten, and the bad things live on? Or is it the opposite? In our family, I only hear good of the dead. I only know of Thatha as a hero, a freedom fighter. The same goes for Gandhi.

I learn that Gandhi isn't perfect either. I read the critiques. His debates with Ambedkar are ones that read like tactic and strategy debates we publish on other causes in *Satya*. Gandhi tries to reform Hinduism by focusing on eradicating untouchability. Ambedkar notes, "There will be outcastes as long as there are castes." Gandhi believes the culture is shifting. Ambedkar challenges him, "I see no change. And what's the good of telling me you are ready to suffer with us? If you have to suffer, it means we will have to continue to suffer still more." Gandhi thinks untouchability is on its last legs. Ambedkar replies, "One swallow does not make a summer. You are highly optimistic. But you know the definition of an optimist? An optimist is one who takes the brightest view of other people's sufferings."

What shocks me about Gandhi is learning about how he tests his Brahmacharya by sleeping naked next to young girls, including his niece. He does not think about the violence caused to her to be the prop for his experiments. I read another article by Ajay Skaria about how satyagraha relies on vulnerability. But it only works when it is your own vulnerability, and it does not rely on the vulnerability of others. When it does—satyagraha can be weaponized.

It's always destabilizing when people you admire and love disappoint you. How do you parse the parts you've inherited (biologically or philosophically) from the parts you wish to leave behind? When I think about Thatha and Gandhi, only one has left me an exhaustive written record to pore through and sort what should live on, and what should be interred.

≈

Sayana Uncle writes a memoir while he was in Zambia. *Lonely But Not Alone*, or is it *Alone But Not Lonely*? What keeps him company are words—quotes he by-hearted in childhood and handwrites by memory on 11x17 paper. He also scribbles the definitions of *ombudsman*, *impeach*, and *amanuensis*. He composed this list:

SIX GIFTS FOR MY CHILDREN

i. self-confidence
ii. enthusiasm
iii. compassion
iv. respect
v. resilience—love is the greatest shock absorber
vi. hope

Are these the gifts Thatha and Patti give you all?

~

What are the inheritances of Kallakurichi? Savi Aunty tells her children, "Be bold. Be brave. Be beautiful." Amirth Uncle's daughter, Anjali, sets up a group chat for us cousins. She said when she and Krishna were young, Amirth Uncle threw them in water to learn to swim. She has no fear, "like a true child of Kallakurichi." We talk about our work, about our arts, about the ways we keep busy—that, too, is a Kallakurichi work ethic. She texts our cousins a picture of the two of us, and Cham Cham chimes in—"I guess all us girls have the same Kallakurichi wide smile. I tell you genes show up at the least expected of places."

I think I have your smile. I found a poem your friend Del Rhodes wrote about you. It came to him in a dream, two years after you passed, and he mailed it to Mom. It was called "Always a Smile." I pause on these lines: "In his Eyes Distant Lands / A touch of Kindness from his hands."

Again, we return to *eyes and hands*. These distant lands and kind hands of Kallakurichi.

~~~

In his memoir, Sayana Uncle writes this quote down twice: "The worst misfortune that can happen to an ordinary man is to have an extraordinary father." What was the price of Thatha's activism and his junoon? Was he around much those sixteen years you had with him? Did your love outweigh your fear of him?

In the back issues of *Satya*, I found the Nigerian activist Ken Saro-Wiwa's last words published:

> We all stand before history. I am a man of peace, of ideas. Appalled by the denigrating poverty of my people who live on a richly endowed land, distressed by their political marginalization and economic strangulation, angered by the devastation of their land, their ultimate heritage, anxious to preserve their right to life and to a decent living, and determined to usher to this country as a whole a fair and just democratic system which protects everyone and every ethnic group and gives us all a valid claim to human civilization, I have devoted my intellectual and material resources, my very life, to a cause in which I have total belief and from which I cannot be blackmailed or intimidated.

I think about Ken Saro-Wiwa's junoon for justice, and about Thatha's too.

A few years later, I read an interview with Ken Saro-Wiwa's son in the *New York Times*. "All of us have a choice, to make our children safe in the world or to make the world safe for our children, and there are implications to that," Mr. Wiwa said.

Did Thatha seek to make his children safe in the world or make the world safe for his children?

Savi Aunty tells me, "Children should not be raised in fear." Amirth Uncle gifts me a book, *Born to Win*, about how to raise successful children by showering them with praise and love. You, too, want for us a gentler childhood.

This is what I think your extraordinary mother offered to compensate for your extraordinary father. She is described to me as all virtue, a bundle of affection wrapped in nine yards of saree. Beneath the layers of fabric is a woman trying to raise a half a dozen of her thirteen children alone, without Thatha and without Gandhi.

―――

In 2009, I tell Amirth Uncle that Wan and I will be visiting India in the summer.

"Wonderful," he says. "We should go to Pondicherry."

"Can you take me to Kallakurichi?" I ask.

He shifts his head backward and pauses for a moment to think. He is not expecting that request. Perhaps it is unfair of me to even ask. But I want to see your birthplace, and I don't know how else to find it, or who else can take me.

"It's not the same. There's nothing there anymore," Amirth Uncle says. "Going there will only make you sad."

"I've never been there. I just want to know where it is and what it's like now," I tell him.

He looks at me, the youngest child of his youngest brother. I can see he, too, sees your face in mine. *Adi's moonji. Adi ditto.* What goes through his head at that moment? Subramaniam Uncle sold the property twenty years prior. You and the rest of your siblings left Kallakurichi long before that. Going back would be a reminder of things he wished to forget, things he lost, the things you lost too.

But he also knows what I've lost. *How can I say no?* perhaps he thinks. You are gone. Who else would take me?

"Okay. If you want to go, I will take you," he says.

He tells Sarojini Aunty the plan. "She'll be in India in July," he says.

"Wonderful! We should all go to Pondicherry," she says. They both really love to visit this seaside city and walk along the boardwalk at night.

"She wants to go to Kallakurichi," he says.

"Why would she want to go there?" she says. "What's the point? There's nothing to see now."

The two talk over each other. I dedicate an ear to each of their arguments.

I meet the two of them for the first time in Chennai—which will always be Madras to you—in 2004, after immersing your ashes in the Ganga. Jaya Aunty warned them I was a vegan environmentalist and told them not to put any ghee or yogurt in my food. Amirth Uncle picks us up at the train station wearing pants that appear too big for him.

"I wasn't sure if my belt was leather or plastic and thought both might offend you, so I didn't wear one," he says upon meeting me. His thick white handlebar mustache turned upward like a second smile.

"Uncle is a show-off," Sarojini Aunty interjects. "He tries hard to impress."

Uncle fries poori and cooks up some alu sabji for us. He teaches me to put a slice of lemon in my water, for its alkaline and anti-cancer properties.

His voice, his laughter, his sense of humor, reminds me of you. I want to see what he looks like without the mustache. *Adi's moonji?* Looking at him makes me feel like you are there, but was also a reminder that you aren't. Perhaps they feel the same when they look at me.

"She wants to know about the ancestral home. Let her go and see for herself," he insists, masking his own reservations. After a few more minutes of debate, they concede to my wish.

They agree to take me to Kallakurichi.

A professor I worked with in graduate school summarizes geology in one sentence: "The tip of Mount Everest is a glacial marine clay." Geology tells the story of how the very bottom of our oceans rise to the highest peak on the planet. I like thinking about tracing origins, feeling like a grain of sand blown by wind and water across the globe. By going to Kallakurichi, would I feel connected to my *parent rock*? That's what they called it in geology.

---

Wan, Mom, and I take a train from Bangalore to Chennai. We are wakened by the singing of little children excited by our arrival in Chennai. Chennai vandhuduthu. Chennai vandhuduthu. *Chennai has arrived. Chennai has arrived.*

We meet Amirth Uncle and Sarojini Aunty there, before heading to Kallakurichi the next morning. Mom wants to avoid the astrologically inauspicious time of Rao Kalum, so we have to leave before seven a.m. She follows these superstitions even if she can't fully explain them to me. Sarojini Aunty has her reservations already about this trip. No good would come from opening past wounds. Why upset the planets too?

While departing their home in Chennai, Uncle recites a Sanskrit shloka wishing us well on this trip. We aren't on the road very long before we make our first stop—a temple, to pray. Uncle is an Air Force Man, so everything about the trip was perfectly timed and scheduled, even the connections with the divine. It is here where I see a temple elephant, chained and confined to the temple grounds. I wish you were here with me, as I attempt to talk with her handler in Tamil.

Next stop is breakfast. Uncle shares his struggles of remaining vegetarian in his Indian Air Force days. He and Aunty were stationed in

Kashmir, where even some of the most devout Hindus ate meat. His fellow officers would laugh when they saw him eating nothing but salad. They remind him of his other impurities. "You drink like a fish, and smoke like a chimney, and you are telling us you are a vegetarian?" He responds that there was no meat in cigarettes or alcohol. I wondered, too, about this radical shift from the nonviolence of Kallakurichi to the violence of Kashmir.

We pull over at a roadside rest stop and have fluffy white idlis dipped in coconut chutney and sambar. Uncle orders dosa to share with me. We sip fresh-squeezed lime juice in stainless steel tumblers.

In the car, Aunty and Uncle resume telling stories at the same time, and I try to pay attention to both. We talk about dreams. Uncle says that Thatha has a dream of Gandhi's death the night before he's assassinated. He doesn't know where Gandhi is at the time, but Thatha writes a letter to the chief minister of Madras to send as a telegraph. Uncle remembers reading that letter.

I ask where those correspondences are now. Uncle doesn't know. He points me to Subramaniam Uncle, who sold the property. When I meet him a few weeks later, he doesn't know and thinks Sundaresan Uncle has them. At the time, Sundaresan Uncle was embracing the sanyasi stage of life, where one lets go of all attachments. He didn't want to talk about the past and had gone into his own exile.

While riding in the car, Sarojini Aunty asks us, "Have you ever had coconut water straight from the shell?" There is a man by the side of the road with a bundle of coconuts and a machete. We pull over and step out of the van. There are discarded coconut shells on the side of the road and a family of goats foraging inside of them. Uncle buys us all coconuts. We drink the coconut water with a straw, and when we are finished, the vendor slices the coconut open with the machete and carves a little coconut shell scoop for us to dig out the insides. It reminds Amirth Uncle of Kallakurichi and the coconut trees he planted there. As children, he says, you all would eat the coconut flesh, moving it around your mouths with your tongues to increase your linguistic flexibility. It was an exercise that improves your Sanskrit pronunciation.

I think about my wedding the year before. Had you been alive, I might have eloped, not wanting a big fuss. But Wan and I wanted to bring the families together. Wan's parents came from Korea and read a Buddhist prayer.

In your absence, I tried to connect with the remaining children of Kallakurichi. Each year brought more losses. But at our wedding, Savi Aunty and Amirth Uncle reunited after not seeing each other for forty years since Patti's death. Jaya Aunty came a month early to stitch saree blouses for me and the other women. Since you were not there, I walked down the aisle carrying a coconut with Mookie, who wears a dress Jaya Aunty made for her.

Amirth Uncle recited a Sanskrit prayer at our wedding. I learned that when he was young, a priest tried to recruit him. Patti intervened, saying, "My husband was a saint. I don't want my son to become one too."

At the wedding, Uncle's Sanskrit praying is a kind of divination, involving singing, exhaling, and concentration. It sounds like you. Savi Aunty says it reminds her of Thatha. Jaya Aunty jumps out of her seat and bows on the grass, paying her elder brother respect. There is a resonance in the syllables. I don't understand what he is chanting, but I am almost in tears. Three generations connected.

I imagine this is what I would feel once more, in Kallakurichi—how the memories of others somehow become mine.

When we reach Kallakurichi, we drive over the bridge Thatha built. Uma Aunty says there used to be a sign with Thatha's name on it, but we find no such marking. A carcass of a goat is rotting by the side of the road.

At the entrance to the property is a big sign curving along an archway: Red Sun Gardens. *So much depends upon* a red sun. There is an ancient sailor's saying, "A red sky at night, sailor's delight. Red sky in the morning, sailor's warning." This red sun in the afternoon is the last glimmer of Thatha's Kallakurichi Ashram, of your childhood home, before it transforms into a new housing complex. We are here with two of the last few souls, along with the Gomukhi river, who remember what this place used to be.

An emaciated, tethered brown cow grazes on the few patches of green grass left. Most of the earth is dry and the vegetation brown. The groves of trees are no longer. Amirth Uncle says that he planted 640 coconut trees. Now, only one remains. We move to the back of the property, where there is a short stone wall.

"Just past that wall, is the river," Uncle says.

We peer over the wall, but see a dried-out riverbed. Amirth Uncle and Sarojini Aunty are saddened once more. Two years back, when they last visited, there was still some greenery and farmland. There was water in the river. The big office was still there, back then. Now, a pile of bricks lay near the entrance. One of the workers at the property says it was demolished six months ago.

The land is barren and crunchy, and parcels are staked out for future homes. Aunty recites a Percy Bysshe Shelley poem.

We look before and after,
And pine for what is not:
Our sincerest laughter
With some pain is fraught;
Our sweetest songs are those that tell of saddest thought.

Sarojini Aunty tells me she lived in Kallakurichi briefly after marriage. She was a city girl from Delhi who never saw a well in her life. In the Gandhian tradition, each member of the Kallakurichi household had to draw their own water. She tells me you made a deal with her. You fetched her water if she played cards with you.

Outside there were three latrines. "One facing the north, one facing the east, one facing the west," Jaya Aunty told me. It wasn't until Sarojini Aunty moved in that they put up a curtain between them. "We didn't know we needed a curtain."

When I tell Sarojini Aunty that Amirth Uncle reminds me of you, she disagrees. "Adi was so sensitive," she says. She said you were different from the others. You couldn't bear to be around bickering, fighting, suffering. In the middle of large family dramas, you quietly disappeared and walked to the bend in the river, away from everyone.

I walked to this place where you once found solace. This hairpin bend in the river is a suitable spot for an elephant birth. Patti delivers five of her children here, including you.

Amirth Uncle and I sit on the ledge of the bridge Thatha built across a dried-out riverbed; only brown alluvial sands and gravels remain.

The Gomukhi means *cow's mouth*. But like the tethered, ribby brown cow on the Kallakurichi land, not much is entering her mouth.

---

"A dying tongue speaks the truth."

Nine months before she died, Savi Aunty tells me stories she never told me before. She was born in 1930 and was a baby when they left Rangoon by boat. Patti narrated these stories to her. Patti, the keeper of stories and memory, gone before I was born. Savi says they arrived in Kallakurichi in 1935.

When Savi Aunty is five or six years old, Thatha takes her to meet Gandhi, and tells him she will be a nurse. Savi protests and asks, why? Gandhi recites a Shamal Bhatt poem to her, which she recites to me, and tells me she wrote on her immigration test to the United States:

A bowl of water give a goodly meal;
A kindly greeting pay them back with zeal.
A tender service 10-40 their reward;
A single penny, pay them back with gold.

Years later, when Savi Aunty was stationed as a nurse at a TB sanatorium in Patna, she had a premonition during her morning ablutions. She requested ten-day casual leave. Upon arriving in Kallakurichi, she gave Thatha a glass of water. "How did you know I was thirsty?" he asked.

"A dying tongue speaks the truth," she tells me.

~~~

Thatha has a dream he is going to die, just like you do. When he wakes, he walks outside and chops the firewood for his own cremation. Does he know in his dream there would be a storm and that the Gomukhi River would overflow? In the days following his death, the Kallakurichi Ashram he created and the road bridge he built floods. The brimming river carrying his ashes pours over his life's work. *For men may come and men may go, but I go on forever.*

# RETURNS

I don't want this to be the end of this story. I don't stop searching for Kallakurichi after I find it.

I learn that the Tamil Sangam poets lived in cities now submerged by the sea. How did we find their poetry underwater?

I learn that Ashoka's edicts, etched in stone, become a mystery that is only unraveled many centuries later.

I learn that, when Gandhi turns to the spinning wheel, the British have nearly eradicated local knowledge of spinning and weaving. That Gandhi asked a widow "to wander from place to place in Gujarat and not rest content till she had found those sisters, who still had the art of hand-spinning in their possession." In Vijapur, she finds a few Muslim sisters who could teach this skill. "From that moment began the great revival which is now covering over fifteen hundred villages in India."

I take a Gandhi Charkha Spinning class in Brooklyn with my friend Beth, the cofounder and publisher of *Satya*. Yukako, the instructor, gathered cotton from the insides of mattresses to recycle for our lessons. The cotton is bunched into stick-like punis that look like cotton candy. We use a portable book charkha—two wheels, a handle for winding, a spindle for spinning. This charkha is the result of a design contest announced by Gandhi, to make a version accessible for the masses (and portable to take to jail). I wonder if this is the kind you used in Kallakurichi.

I loosen the puni, so I can catch the short fibers onto the spindle with my left hand while my right hand turns the wheel clockwise. The left hand pulls back in draft with a soothing *knock knock knock*, when done correctly, like Yukako demonstrates, and the fiber turns to thread slipping through the fingers. Unwind a little bit to push the thread on the spindle back and wind again. It is much harder than it looks. And I am in awe of this being a strategy for the masses to reclaim independence.

Although I have read the history of spinning, I do not fully appreciate it until I am sitting on the floor trying to find the rhythm, the angle, the speed, the grip, the patience, and the faith, that made the rotations a revolution. To reclaim a knowledge that has been forcefully eradicated by the powers of empire. To imagine the possibilities of a nonviolent future in a spool of thread.

I am interested in learning how to find what is nearly lost, to revive what was thought to have been erased.

～～

"Vanakkam," my Tamil tutor, Sasipriya, says, and brings both hands together in prayer. "We do this when we greet each other, because we see God in everyone." I take Tamil lessons, another attempt at reviving what is nearly lost.

I tell Sasipriya, who lives near Madurai, about you and Thatha and Kallakurichi. I say I am interested in the history of satya and ahimsa. She wants to help me, but tells me India has changed. "There is no satya, no ahimsa." India today is not what Thatha dreamed of, or what you left in Kallakurichi.

Sasipriya recited a Tamil expression. Katrathu kai man alavu. Kalaathathu ulagalavu. *What I know is in the fistful of sand in my hand. What I don't know is the whole ocean.* She will share whatever she knows, like the three-striped squirrel who helped Rama grout the holes between the stones in his bridge to Lanka.

～

I keep searching for Kallakurichi in the lacunae of archives. On my first day in the Shoichi Noma research room of the New York Public Library, I am greeted by the stone lions, Patience and Fortitude. I request two decades of Gandhi's final newspaper project from the underground stacks. Eighteen purple, bound volumes are delivered to my room and stored on my shelf. Since I do not have Thatha's writings or records, I find a proxy in Gandhi's newspaper articles, which often read like instruction manuals.

I turn to Italo Calvino's *Invisible Cities*, an imagined conversation between Marco Polo and Kublai Khan. "Arriving at each new city, the traveler finds again a past of his that he did not know he had," Calvino writes.

I understand this as desire—a yearning to know an unknown history.

"Sometimes different cities follow one another on the same site and under the same name, born and dying without knowing one another, without communication among themselves," Calvino continues.

I think about this as I sit at this library that was built on the site of a former drinking water reservoir. I am reading about the submerged cities of the Sangam poets, from texts stored below the skating rink in Bryant Park and delivered to me via an underground trolley system, a portal connecting one city on top of another city.

The early Tamil Sangam literature is lost to katalkol—*seizure by the sea*.

The poets move inland, and the sea follows. After each city submerges, a new one emerges bearing the same name—Madurai—the amphibious kingdom of our ancestors.

Do their poems sink or swim?

My Kinyarwanda translator T told me that there are similarities in the philosophies of Dravidian and African languages. We talked about my name and similar words in African languages. "In Kinyarwanda, *nsanga* means *I find*, or *I found*," T told me. "In Lingala, s*ango*, or *sangu* is in the sentence, *Sango nini?* Meaning, *what's the news?* Or, *how are you?* Others will say *Nsangu nini?*

"The answer is *sango*, or *nsango malamu. Good news*, or *I am fine.*"

I look up the book T recommended and learn the legend of Kumari Kandam, the submerged continent linking India to Africa. A British zoologist called it Lemuria to hypothesize how the ring-tailed, wet-nosed, leaf-eating prosimians traveled across oceans.

Can we imagine these epic journeys of lemurs and languages?

You once took me to the tip of India, Kanyakumari. We were on the verge of this lost land bridge.

"This is a book about place-making and imagining," Sumathi Ramaswamy writes in her book *The Lost Land of Lemuria: Fabulous Geographies, Catastrophic Histories*. "It is a lost place from a lost time." I think about my own attempts at place-making with Kallakurichi, another lost place from a lost time, and how this journey has led me to this lost subcontinent of our submerged ancestors, of other stories of hidden and forgotten homelands, and the lost knowledges (real or imagined) contained within.

# CATENA

As a teenager, I stumble upon a list of famous vegetarians. It's in the *Vegetarian Times* magazine in the waiting room at my chiropractor's office. I feel a connection to them and am curious why we were never taught this about them. In school, we study Leonardo da Vinci's artwork but are never told he went to the live markets to buy birds to free them.

$$a^2 + b^2 = c^2$$

The Pythagorean theorem is instilled in us, but not the equally indelible stories of Pythagoras saving dogs and fish, or that vegetarians during his time were called *Pythagoreans*. Decades later, while reading Colin Spencer's *The Heretic's Feast*, I learn that the Egyptian priests, with whom Pythagoras studied as a young man, are likely his entry point into vegetarian eating and thinking (and probably the Pythagorean theorem too). "Egyptian priests were also particular about not wearing any clothing that was derived from animals. Wool was banned, even as a shroud to be burned in; their clothes were made from linen and their sandals from papyrus," Spencer writes. I love learning about all these lineages.

At the New York Public Library, I unbox a fragile volume of Howard Williams's *The Ethics of Diet*, published in the late nineteenth century. It is the subtitle that catches my interest: "A Catena of Authorities Deprecatory of the Practice of Flesh-Eating."

A *catena*. A chain of linked texts.

Williams compiles an anthology of thinkers who critiqued the consumption of animals, linking Prince Siddhartha to King Ashoka to Pythagoras to Plato. Williams goes on to assemble over forty of their descendants in vegetarian thought. Gandhi read it during his London days, and his literary hero Tolstoy, who in true Tolstoy fashion penned a forty-plus-pages introduction to the Russian edition.

In his essay introducing the catena, Tolstoy writes: "We cannot pretend that we do not know this. We are not ostriches, and cannot believe that if we refuse to look at what we do not wish to see, it will not exist. This is especially the case when what we do not wish to see is what we wish to eat."

In soil science, a *catena* is a sequence of distinct soil layers that coevolve down on a slope. Each are unique but connected. In my journeys to recover ancestral knowledge, I realize there were other lineages I should be exploring, too—lineages of ideas, of influences. Sometimes we know our forebearers in philosophy, and sometimes, like Calvino's cities, certain thoughts follow one another under a similar name, at different times, without knowing one another, without communicating among themselves.

~~~

When I work at *Satya*, we record interviews on microcassettes. The recorder has an attachment we connect to a landline phone. We use this transcription machine with a foot pedal, and you never have to remove your hand from the keyboard. I would stop, rewind, and play, over and over, to fully understand what was said and meant. Each time, new meaning, understanding, or questions arise. Accruing and arranging until I find meaning.

# VERGE

There's humility in witnessing earth being created. Orange fire and new black rock expand to meet blue waves. The elements hold so much power, they could be mistaken for violence. As I sit at the edge of the cliff off the Big Island in Hawaii where these elements collide, my body healing from a recent flare-up, the trees bending toward the rising sun as if performing Surya Namaskaram, a sea turtle bobs on the ocean surface below.

I am at a writing retreat where the brilliant poet Bhanu Kapil gives us instructions from afar via a video chat: *Walk until you find a verge. Lay down in the verge. Take notes in the verge. Return to a verge in your work. Transport these verge notes to the verge in your work.*

Jaya Aunty tells me you are born a poet.

Is this how a poet's mind works, carrying meaning between experiences, giving language to language?

In her book, *The Vertical Interrogation of Strangers*, Kapil interviews women of South Asian descent and asks them to answer one of twelve questions. Question six is the one that hits me in the gut: "Who is responsible for the suffering of your mother?"

Bhanu does not ask, *Did your mother suffer?* That is a given.

Do you know who was responsible for the suffering of your mother, the suffering of Patti?

Is it Gandhi? Patti said, "Gandhi came in my youth, and when he left, my youth was gone."

Is it Thatha? Patti said, "My husband was a saint. I don't want my son to become one too."

Bhanu does not ask, *who is responsible for the suffering of your father?* But I do.

Do you think your suffering was responsible for the suffering of my mother?

※

When you leave the earthly plane, Mom's heart starts to palpitate, and she returns to the cardiologist she first saw seventeen years prior when her father passed away. Do you remember, back then, how only one person had a phone on Narayanappa Layout in Bangalore? I can picture the other end of the line, our red phone in the kitchen that told Mom that Thatha had died. It was 1986 and Mom took the first flight back to India, while you were in charge of Manu and me. I was in third grade, and he in fifth. You did my hair, wrapping those marble-ended bands, giving me lopsided pigtails.

The day she departed was the same day of the Challenger explosion. The next morning at school everyone was in mourning, and Manu and I were so sad our mom was away. During her absence, I slept hugging a pillow, my Mom pillow.

Mom's heart was breaking then, too, and these heart pains return to her when you pass away.

In the years that pass she finds community.

She joins an elders organization, Jeevan Jyothi, which means *light of life*. Your friends Mamta and Suresh founded this Indo-American support group for seniors. Remember, Mamta was pregnant at the same time Mom was pregnant with me. They swapped food, when they each craved each other's cooking during pregnancy instead of their own.

Now Mom also serves on the board of this organization, and facilitates book club, yoga, and Bollywood-themed Sunday night sing-alongs (her version of SNL). They plan trips together. She has visited Argentina, Brazil, Chile, China, Ecuador, Greece, Russia, and Turkey. She and the other aunties carry snacks and spices in their bags, to share with one another. Mom is known for carrying everything—hand sanitizer, medicine, snacks, ready to dispense.

Nearly two decades after you leave us, the Rockland County Legislature honors Mom for Indian Heritage Month. They note her work as a social worker for thirty-seven years, her involvement with Jeevan Jyothi, and her many invisible kindnesses: how she takes members to doctors appointments; cooks for others when they are sick; and stocks their fridges before they return from trips.

In her acceptance speech, Mom, always humble, recognized others who, like her, "quietly do the work." The next day she travels to India to be there for Varalakshmi Nombu.

I learn Varalakshmi Nombu is a festival to bring health, wealth, and prosperity, performed by married women, to pray to their goddess for the wellbeing of their loved ones. I did this once in India, the year after my wedding, not knowing, understanding, or appreciating much. My maternal Patti tied my nine yards of orange and purple. The menfolk read the newspaper. "We can sit this one out," they told Wan.

I think of Lakshmi: the name of both of my Pattis; my middle name; the goddess of wealth, sitting on a lotus flower, four arms like tentacles, each an offering, gold coins shooting out of palms, the granter of boons. I think of all those women praying to Lakshmi. Do they not see their own arms, always extended in offering, blessing, and caring? Do they realize they were the goddesses of boons all along?

I am arriving at some grand, unifying, mind-gut, divining poetry, vegan theory. My friend Leslie reminds me to define my terms. My friend Iselin sends me a tote bag that reads: "Not *plant-based* as in kale chips, *vegan*, as in overthrow the patriarchy."

I'm looking to define a vegan poetics that extends beyond diet, beyond words—a way of being, living, holding, bearing, connecting, transforming.

You are a poet. You introduce me to poetry. I recite poetry to you as you are dying, not knowing you are dying. But we never discuss what poetry is. It is unspoken and understood, perhaps like the suffering of mothers.

In an interview with *The Black Scholar* journal, James Baldwin speaks about poetry beyond the lens of literature. He talks about poets as being able to recreate an experience to give it back to you, to make it bearable. "And if you could bear it, then, you could begin to change it," he says.

Maybe poetry is like floating on your back in quicksand—the bodily knowledge that saves you.

---

In high school, I read "The Red Wheelbarrow" by William Carlos Williams.

> so much depends
> upon
>
> a red wheel
> barrow
>
> glazed with rain
> water

beside the white
chickens

My AP English teacher, Mr. Spaight, focuses on the word *depends*, meaning to hang from, just like how all the words are suspended vertically from the word *depends*. The poem hangs from it. "It's not *what* a poem, means," Mr. Spaight says. "It's *how* a poem means."

I have always been fascinated by the question of *how*. How to live, learn, care, grieve, love, remember, forget, forgive, imagine. I have been looking for meaning in process—in searching, finding, losing, disassembling, and reassembling.

William Carlos Williams also writes that "It is difficult / to get the news from poems / yet men die miserably every day / for lack / of what is found there."

I have been poring through Gandhi's newspapers from the 1930s, searching for Thatha, for history, for memory, for Kallakurichi, an invisible city. I am not looking for the news in poems, but poems in the news.

It's not long before I find a poet distilling lifesaving bodily knowledge in these pages. Rabindranath Tagore penned a passionate plea in 1936 titled "Self-Imposed Impoverishment" warning of industrialized rice mills, which remove vital nutrients while polishing rice.

"When a people's diet takes a vicious path of its own impoverishment it causes a graver mischief than any act of cruelty inflicted by an alien power. [. . .] Rice mills are menacingly spreading fast extending through the province an unholy alliance with malaria and other flag-bearers of death robbing the whole people of its vitality through a constant weakening of its nourishment."

The newspaper readers who respond posing questions are called *posers*. One poser writes Gandhi, "Don't you think that it is impossible to achieve any great reform without winning political power? The present economic structure has also got to be tackled.

No reconstruction is possible without a political reconstruction, and I am afraid all this talk of polished and unpolished rice, balanced diet and so on and so forth is mere moonshine."

I know this poser, or his descendants. *What is the point of veganism, poetry, or doing anything?* they may ask me decades later, not working on political reform either.

Gandhi knows him too: "I have often heard this argument advanced as an excuse for failure to do many things," he replies. "I admit that there are certain things which cannot be done without political power, but there are numerous other things which do not at all depend upon political power."

"The correspondent has raised the bugbear of 'great' reform and then fought shy of it," Gandhi continues. "He who is not ready for small reforms will never be ready for great reforms."

This reminds me of what the activist adrienne maree brown said in conversation with David Naimon about something she calls *fractal responsibility*. "If we are thinking that we are going to be able to change something at the large scale, but we have no embodied practice with on a small scale, we are operating upon a delusion."

It makes me think about what Jane Addams wrote in the introduction to the English edition of Tolstoy's *What Then Must We Do?*, which Gandhi also read:

> It may be no clearer to us than it was to him [Tolstoy] that a righteous life cannot be lived in a society that is not righteous. It was clearer to him than it has been to any others, save to a small handful of shining souls, that the true man can attempt nothing less and that society can be made righteous in no other way.

Addams also wrote that "Realizing also as we grow older that life can never be logical and consistent, it still remains the fact that Tolstoy makes complacency as impossible now as when *What Then Must We Do?* first appeared."

Gandhi, too, makes complacency impossible and finds agency and power in daily life—work that the masses can undertake to improve their lives and possibly save them.

"That is why I can take the keenest interest in discussing vitamins and leafy vegetables and unpolished rice," Gandhi writes. "That is why it has become a matter of absorbing interest to me to find out how best to clean our latrines, how best to save our people from the heinous sin of fouling Mother Earth every morning."

My friend asks me if I thought Gandhi was a poet. I pause for a moment because it's not the usual descriptor. After spending more time with his words, I see him as a serial diarist, accounting for almost every thought and action, and how these may have shifted over time. When I look to these newspaper projects, I appreciate Gandhi's role as editor, the connections he makes with seemingly disparate topics, the synergy between the individual and the society, the juxtaposition and inclusion of literary excerpts. I see echoes in *Satya*. While we weren't a poetry journal either, we dealt with these similar linkages and made ourselves vulnerable to truth.

There are many poetic descriptions in Gandhi's language that are indelible. The bleating goat in his stomach, how he described partition as "vivisection," and that feeling of spotting a vegetarian restaurant in London, as "a child getting a thing after his own heart." Yes, I think he was a poet.

I see the poetics in the coining of *satyagraha*—finding language for the thing to hold all things. He chose truth as the anchor instead of any cause or action to unify all his causes and actions. Ahimsa, too, was applied to everything.

"Non-violence is at the root of every one of my activities and therefore also of the three public activities on which I am just now visibly concentrating all my energy," Gandhi writes in an article titled, "The Greatest Force," in the October 12, 1935, edition of the newspaper. "These are [the eradication of] untouchability, Khadi, and village regeneration in general. Hindu Muslim unity is my fourth love." He further notes, "I cannot be non-violent about one activity of mine and violent about others. That would be a policy and not a life force."

It is this life force that I am looking for in this newspaper, in *Satya*, and in life.

When I think of Martin and Beth starting a publication that linked vegetarianism with environmentalism, animal advocacy, and social justice, they, too, come to truth as the unifying tenant. Just like truth was the first step in Tolstoy's *What Then Must We Do?* "First: not to lie to myself; and—however far my path of life may be from the true path disclosed by my reason—not to fear the truth." This also reminds me of Jaya Aunty's adage: "Fear not the path of truth for the lack of people walking on it."

---

The Syrian poet Kahlil Gibran makes an appearance in Gandhi's newspaper too. A translated excerpt of his poem "The Prophet" is reprinted under a section called "The Glory of Work," in the November 23, 1935, issue: "You work that you may keep pace with the earth and the soul of the earth."

It is in this next line that I hear the echoes of Jaya Aunty and Savi Aunty's gospels of work: "Work is love made visible."

If there is a satyagrahi poetics, this could be its mantra. Or perhaps, *"Love is work made visible."* The newspaper made the work of swaraj visible: spinning cotton, siting latrines, digging wells, opening temples, eating leafy vegetables, administering first aid. These are the poetics of work.

While Thoreau read the Gita in his Walden retreat, Gandhi tells his new recruits: "Go not to the villages to write commentaries on the Upanishads, which you may well do in towns and cities. Your work will provide the best commentary."

"Infect the villagers with our passion for work," he says.

What I discover in the poetry of the news is that a snake can also be a satyagrahi. "As co-workers are spreading out in villages, it becomes necessary to arm them with information in matters of common occurrence in villages where city conveniences are unfortunately unavailable," Gandhi writes about the challenges of village rejuvenation. "The most dangerous is a snake bite which proves fatal in many cases if the necessary measures are not immediately adopted." In 1935, a sadhu offers a new remedy that requires test subjects and "Gandhiji, in his quenchless thirst for knowledge, and ready for any new experiment, provided it could help him to be better equipped for the service of the poor, readily agreed." This is met with unanimous protest. You can imagine the roaring and the fainting that occurred at the mere mention of the thought of Gandhiji being harmed. Two of his friends volunteer in his stead, "but the snake proved too Satyagrahi to be coaxed to inject his venom into the blood of the ready victim."

This anecdote charms me. Gandhi says that nine-tenths of the snakes are nonpoisonous and it's important to distinguish the poisonous from nonpoisonous. These were the lessons of your childhood too. Natural and easily accessible remedies are circulated in *Harijan*, like the one Gandhi relates from Adolf Just's book *Return to Nature!*: "Take as much clean earth as possible, add cold water, make a cold poultice."

The earth will save you.

Was Thatha reading all these remedies when he decides to write Gandhi with his own offering? I was hoping to find Thatha's treatise on divining water in these archives. What I discover instead, one January afternoon, sitting in the Noma Room of the NYPL after a morning swim, while flipping through a purple bound volume of articles from 1935 and landing on the October 19, 1935, issue of *Harijan*, is this:

## SIMPLE CURE FOR SCORPION STINGS

Shri M. S. Narayanan who is a retired engineer writes the following letter:

Scorpion stings are very frequent here and I am called upon to attend upto a dozen cases a day. I tried known remedies and there were no complete successes. I bought years ago some phials advertised by some 'Himalayan remedies' company. A drop had to be put into each eye and the patient immediately smiled and said he was all right. The phials had got exhausted except one the label had been eaten away by insects, and so I did not know where to get the phials from again. Each cost 8 annas and contained about 15 drops. Some weeks ago I came across a sadhu who removed my distress by suggesting a very common remedy. This is to shake common salt in a bottle, to make a saturated solution, to allow it to settle for a day and decant the clean solution and keep. A drop of this solution put into each eye removes the pain of the sting instantaneously. I have been trying this remedy for the past four months and have treated 276 cases in all of which there has been complete success. In 270 cases the cure was instantaneous. In 6 cases the pain decreased gradually. Three applications were necessary to bring the pain to the vanishing point. The last phial which I had was put to chemical test and revealed pure Na. Cl. I did not want to write to you till I had tried a good number of cases. Maganwadi has a decent number of scorpions and after local tests you may give the information to the rest of the world.

### Simple Cure for Scorpion Stings

Shri M. S. Narayanan who is a retired engineer writes the following letter:

"Scorpion stings are very frequent here and I am called upon to attend upto a dozen cases a day. I tried known remedies and there were no complete successes. I bought years ago some phials advertised by some 'Himalayan remedies' company. A drop had to be put into each eye and the patient immediately smiled and said he was all right. The phials had got exhausted except one and the label had been eaten away by insects, and so I did not know where to get the phials from again. Each cost 8 annas and contained about 15 drops. Some weeks ago I came across a sadhu who removed my distress by suggesting a very common remedy. This is to shake common salt in a bottle, to make a saturated solution, to allow it to settle for a day and decant the clean solution and keep. A drop of this solution put into each eye removes the pain of the sting instantaneously. I have been trying this remedy for the past four months and have treated 276 cases in all of which there has been complete success. In 270 cases the cure was instantaneous. In 6 cases the pain decreased gradually. Three applications were necessary to bring the pain to the vanishing point. The last phial which I had was put to chemical test and revealed pure Na. Cl. I did not want to write to you till I had tried a good number of cases. Maganwadi has a decent number of scorpions and after local tests you may give the information to the rest of the world."

The caution Shri Narayanan gives is worthy of him. He is devoting his savings and his time and skill to village reconstruction principally through khadi. Though Maganwadi has a fair share of scorpions, the cases are nothing so frequent as Shri Narayanan has. I have no reason to doubt his testimony. I must not therefore keep this costless remedy from the public for fear of its proving a failure. Those who will try the cure will please report the results of their observations. If I have reports only of failures, the public shall know them. Let those who will try the cure prepare the solution in the manner prescribed.

M. K. G.

These 266 words are all that I've found of your father's writings. I reread, repeat, and retype them to try to divine his voice. Is this what Thatha sounded like? Is this his poetics? Part of the work of village rejuvenation was figuring out coexistence with other beings. Jeevakarunyam, *compassion to all living beings*—a Kallakurichi poetics.

---

I learn about the colonies of bees at Gandhi's ashram. *Harijan* prints what looks like a bee family tree. "The bees have added real joy to life here," C. Rajagopalachari writes in an article, "The Story of Our Bees," in the January 25, 1936, issue of *Harijan*. "More precious than honey are the things we gain from the company of these wonderful little creatures. To learn to deal with them is a continual lesson in the great art of love, gentleness and consideration, an education of heart, mind and hand, and process of true civilization."

The newspaper talks of the differences between apiculture and the wild-honey gatherer. "The latter does nothing for the bees. He destroys the combs and brood when extracting honey. The surviving bees if any must build a hive anew somewhere. The honey-gatherer becomes thus one of the deadliest enemies of the species."

In apiculture, the columnist argues, "the bees get comfortable and well protected hives and have all the advantages of their natural home without its dangers. [. . .] Let there be no mistake. There is in my opinion no virtue in keeping bees. Beekeeping is unadulterated imperialism. To do without honey is best."

I get drawn into the evolving drama of the bees, which span several issues. I wonder if the bees at the Gandhi ashram will ever achieve their swaraj?

There is much focus in *Harijan* on insects and other smaller creatures. One article is about vermin-proofing paper to prevent infestation from silverfish. Arsenic is suggested to protect the paper. A reader named Rajaji writes in requesting a "reprieve from Socratic doom you have condemned to the silverfish" noting how wrong it is to "look upon this bright little insect as an enemy to civilization."

The reader disputes the notion that this poisoned paper is better quality. Even if so, it's not a good enough reason to use it. "Dead hide need not be actually superior to the slaughtered one for us to support the former!"

I appreciate this, too, where the seemingly "better" choice is not necessarily the best one, or the "best" choice is not the better one. It depends on the lens.

～

I publish an essay where I talk about you and Thatha. I write about divining: "All I needed was an L- or Y-shaped branch, and someone to teach me."

Another writer I haven't yet met, Shruti, reads the essay and sends it to her father, Sanjay, who writes me an email:

> I am really moved by your writing and your story. I am probably around your dad's age. I lived for two decades in the Santa Cruz Mountains, where Shruti spent her formative years.
>
> One of the many unlikely things I learned while living in nature was dowsing (needless to say, I'm an engineer by profession).
>
> Which is why I'm writing to you.
>
> In memory of your Appa and Thatha, I'll teach you how to dowse.

I don't really expect you to respond to an offer from an internet stranger, but I make it anyway.

Many blessings to you on your path!

---

We have our first dowsing lesson as a video chat a few years later. Sanjay is in California, and I am in New York. We start with philosophy.

"So, the body actually has a mind of its own," Sanjay begins. "It's like we have two minds, we have this conscious awareness, which is who we feel we are, you know, this feeling of 'I am' that is within us. But there is [also] something that runs the body.

"What is this energy that courses through the body, that is our vitality?" Sanjay continues. "'The force that through the green fuse drives the flower'—that's Dylan Thomas," he says, introducing prana, or qi.

"And we live in this society, which has completely discounted the existence of that because we don't have tools that can measure it . . . therefore it doesn't exist or is unscientific," Sanjay says. "But actually, the most exquisite tool for sensing and measuring it is the body."

He tells me there's a paradox, a schism, a chasm "between what our conscious mind is aware of and what the mind of the body is aware of."

"So, what happens in dowsing is that we are going to make a connection between these two minds—the mind that you are using right now to understand me and the mind of the body which knows, and it can't 'English' what it knows. But it really knows how it feels."

This sounds like poetry to me.

I tell him when I hear him talk of two minds, I think about my desire to make connections in my writing following a kind of bodily knowledge. I tell him I am trying to articulate a poetics for it, because I think perhaps "poetry is the place where it can hold it all, where you can make these big leaps that are distilled together."

I tell him I'm working toward finding the life force in my own writing, "the connections that run through it, giving language to it when maybe language is insufficient."

≈≈

At the end of *Invisible Cities*, Khan points to futility if the inferno is where we are all heading. Polo responds: "There are two ways to escape suffering it. The first is easy for many: accept the inferno and become such a part of it that you can no longer see it. The second is risky and demands constant vigilance and apprehension: seek and learn to recognize who and what, in the midst of the inferno, are not inferno, then make them endure, give them space."

I think also about what Sanjay said about divining being a way of connecting our two minds. Choosing not to become the inferno like Calvino says, choosing not to bury our heads in the sand like Tolstoy alludes, choosing to give space to authorities deprecatory of flesh-eating like Howard Williams has done, requires connecting these two minds, paying attention to our inner guides, our inner gut. Return to a verge in your work. These are all the vegan poetics I'm divining. I want to connect to this catena and extend the chain.

≈≈

All my sources are converging. My Tamil teacher, dowsing guru, and Kinyarwanda translator all lead me to the Sangam poets, to Kumari Kandam, and to the translator and poet A. K. Ramanujan. David Shulman credits Ramanujan for providing English translations of these ancient poems that "shimmer and smolder."

I think about how it is even possible that all their words converge on my shelf at the library. The poetry of the first Tamil Sangam era is swallowed by the sea. The grammar *Tolkāppiyam* from the second remains. The third includes eight anthologies and ten long poems. These are written nearly two millennia ago and have been largely lost until the nineteenth century. We owe Swaminatha Aiyar, who encounters the Tamil poets when he is forty-four and decides to spend the rest of his life roaming the villages, monasteries, and private attics to rescue and reprint the salvaged ancient texts.

The results of his labors find their way to the University of Chicago Library, where in 1962, Ramanujan first encounters them as a student.

I find this Ramanujan passage as an epigraph to Vijaya Nagarajan's book on kolams, while sitting in the New York Public Library:

> Even one's own tradition is not one's birthright; it has to be earned, repossessed. The old bards earned it by apprenticing themselves to the masters. One chooses and translates a part of one's past to make it present to oneself and maybe to others. One comes face to face with it sometimes in faraway places, as I did.
> (Ramanujan, 1985)

David Shulman was on the subway one day in New York when he recognized one of the "Poetry in Motion" posters on display as a Ramanujan translation.

"But in love our hearts are as red earth and pouring rain: mingled beyond parting." The poet is named Cempulappeyanirar, which translates as *The Poet of the Red Earth and Pouring Rain*. This sounds like a geotechnical engineering poet to me. Ramanujan notes, "Poets were known by their poems. Their metaphors were their signatures."

What would my Sangam poet name be? What metaphor is my signature? Could it reflect how I think and see the world? Maybe it is the name you have given me already. *Friend of confluences.*

≈

I am asked to contribute to an anthology called *Letters to a New Vegan*, inspired by Rilke's *Letters to a Young Poet*. "Being vegan is more than just what we choose to eat and not eat. In time, you'll come to realize that is the easy part," I write. "The difficult part lies within us—how we find beauty, joy and courage to counter the violence teeming around us."

"You may feel overwhelmed carrying this burden, but I hope you come to realize what a wonderful gift it is to open yourself up to this compassion, and be sure to extend this kindness to yourself . . . I try to find the part of me that had the courage to say, 'I won't' many years ago. As a new vegan, you may not yet have the words to explain your choices, but there is something inside you that is telling you that this is right. As Rilke said, 'go into yourself.' Find that part of you—the best of you—and protect it."

---

I write with my intuition, with my gut. I must also work hard to protect it. Like poetry and ahimsa, my ulcerative colitis is another one of my inheritances from you. "The term 'ulcerative colitis' was first used to describe the disorder by a British pathologist, Samuel Wilks, in 1859," writes Dr. Antonina Mikocka-Walus. Wilks's report, published in the *Medical Times Gazette*, was called "The morbid appearance of the intestine of Miss Banks."

So, we have Miss Banks to thank for a diagnosis of our invisible disease. We are still learning so much about mind-gut connections but there is something about colitis—its vulnerability, the bodily knowledge, that I think is connected to my vegan poetics. Why do we say *gut-wrenching*, or have butterflies in our stomach, or trust our gut?

After my last colonoscopy, I wake from anesthesia and see my nurse out of focus, and then in focus. Words garbled and then clear. My GI returns to tell me my colon looks great top to bottom, deep remission. A relief and a surprise, that my colon could heal. My colon has been the indicator species of my body, the first to know something is wrong, to sound the alarm by attacking itself. In tune with all outside stresses, hypervigilant, hypersensitive. But now it is the resilient bunker when the world is collapsing around it.

~~~

Sanjay talks about energy channels, and chakras. Outside Western culture, "they have very detailed maps of the body." We then discuss structures. Sanjay describes Vastu as Indian feng shui. "If you look at any enclosed space, it has an energy body. Just like we have an energy body. [. . .] It makes for a more beautiful life, if you remove the sources of energetic disharmony in your space." He's visited homes with negative energy and death. By contrast, he tells me about taking his dowsing rods and mapping the energy in his current home. "From deep in the ground there are wellsprings of energy that come and run through the house. And they thread through every room. It's a three-story structure. So, the rooms are like beads on a chain, and it's this beautiful earth energy welling up that's going through the house."

"If a deer would walk through my property, [she] leaves an invisible etheric imprint so I can take this dowsing rod and have it point me to where the deer walked," he also tells me.

I tell him his descriptions remind me of Calvino's *Invisible Cities*, hidden ones with energy rising or falling or moving.

We then talk about geomancy. "If you look at the lands where people have lived or people have died, where people have fought wars, where people have done all kinds of or are being subjected to atrocities of all kinds . . . The earth has a body. The earth is a living being. And this living being has its own energetic structure and its own channels of energy. Just like we do. And they express like veins of energy across the surface of the earth."

I think back to Tara the elephant and what she sensed when she approached the historic battlefield of Kalinga. Would geomancy reveal the sites of istembabwoko in Rwanda?

~~~

I write Shruti and send Sanjay a voice memo.

Hi Sanjay,

It's Sangu. I was just writing to Shruti about how I'm in a whale kick right now, soaking up interesting tidbits of their lives. The combination of water and music also made me think of you.

The male humpback whales are the singers. Aligning themselves vertically, face to ocean bottom, tail to sky, they sing to the newborns to orient them to the world. They sing into the ley lines of the earth. Perhaps they are the original diviners.

In the clear warm waters surrounding the heart-shaped atoll of Moʻorea, perhaps the once heart of Lemuria in what is now French Polynesia, whale mothers come to give birth, and whale fathers come to sing.

Can we imagine bathing in their sounds, floating on our backs, ears in water, feeling their song vibrate from within our own bodies?

≈

Sasipriya, my Tamil tutor, says we use different pronouns for describing location or feeling.
"I is *Naan*, if you want to say, 'I am on the verge.'"
"I is *En*, if you want to say, 'I feel excited.'"
"Add *ku* to *en* if you want to say what you like, love, want, know, or understand."
Enaku enna vennum? *What do I want?*
To know, love, and understand everything submerged.

Water dowsing is the most basic ancient survival skill, Sanjay tells me. But "you can use dowsing to find hidden things, you know, all kinds of lost things, when you are out in the world looking for stuff."

I learn water words as a kind of remembering. David Shulman says one interpretation of the etymology of Tamil is *waterness*.

*water*—thanni
*river*—aaru
*well*—kinaru
*stream*—oadai
*ocean*—samudhiram

The word for *lake* is the same as the word for *tank*—eris. The early Tamils were amazing water engineers, who created a system of eris to store water, allow recharging of groundwater, and slow the flow to prevent erosion and flooding. The British admired these systems but destroyed them when they came to India. Now we look to remember this relationship with water.

I ask Sasipriya how to say *water diviner*.

Neerotam paarpavar—*stream seer*
Neer Kuri Solpavar—*water predictor*
Enoda Thatha—*my grandfather*

---

I keep searching for meaning in Tamil and wisdom in the Tamil poetics. The Tamil treatise *Tolkāppiyam* is a grammar of literature and life. If I want to have the answers to all life's problems, Sasipriya tells me, I must read Tiruvalluvar's *Tirukkural*, which they guesstimate was written between second century BCE and eighth century CE. We don't know much about Tiruvalluvar. Perhaps he was a weaver. Perhaps he was the child of a Brahmin and a Dalit.

Perhaps like Buddha or Mahavir, he was a prince turned spiritual guide. Tiruvalluvar doesn't believe your worth comes from your birth, but rather, how you live your life.

I collect several versions in translation. Mom gives me hers. Nana Thatha buys me another and writes "By reading this book your knowledge will improve," and I find a couple more.

These 1,330 couplets doling out wisdom are a lot to take in at once, so I start with the ones on empathy, compassion, and rain.

This reads like a principle of *Satya*:

> 257:   Know meat for an animal's sore that it is,
>           And you will not eat it.

The second chapter is in praise of rain:

> 20:   No being can be without water—nothing can flow
>          For anyone without rain.

These remind me of you:

> 573:   What use is a raga that cannot be sung?
>           Or eyes without sympathy?

> 575:   The jewel of the eye is sympathy; without it
>           Eyes are but sores.

"Eyes are sympathetic," as you always said.

～

I delight in reading *The Treatment of Nature in Sangam Literature* by M. Varadarajan in my research room at the library. He notes how the Sangam poets wrote about nature without having a separate name for it, how trees are intertwined with the species who inhabit them: "vênkai with the peacock, the mango with the cuckoo [. . .] and the ya with the kite." Varadarajan notes, "The rattan bush is described as the refuge of the otter."

Who were the animals who took refuge in the coconut, lemon, cashew, and pis a pis trees in Kallakurichi?

I wish I could read the poems in Tamil, but through Varadarajan's English explanations I see the pictures the poet Maturaik Kanakkayanar creates of a tiger hiding in the bamboo, who then becomes exposed when the wind blows, or how the vênkai tree with its dots and stripes resembles a tiger. "This fact has attracted the imagination of many poets of the age, and they have described the elephants attacking with vengeance or running away from the tree mistaken for a tiger."

I notice how some Sangam poets portray animal joy and agency.

"A female monkey has clandestine union with a male and trying to escape the notice of its group, looks into a deep pool of clear water, uses it as mirror and sets right [her] hair on the head," Varadarajan writes. I think of you in that first visit with Mom, combing your hair to one side.

The Sangam poets are writing in a world far less polluted and degraded than today. Observe as Varadarajan does, how a poet writes about the pleasures of a monkey who drinks stream water infused with fermented jackfruit and plantain, even though the intoxication will cause her to later fall from a sandalwood tree, "but only on a comfortable bed of fragrant soft flowers," in time to catch the display of a dancing peacock.

In this older Tamil world, the monkeys are so accustomed to beautifully fragrant smells, "that when a flock of cranes comes and perches on the branches of a jack tree, they sneeze, on account of the bad odour of fish that the birds emit."

Varadarajan conveys a scene with the leaves of the plantain tree gently stroking the elephant as she lays near it, and the aerial roots of the irri tree caress the elephant to sleep, while "on the bank of a stream, the noise of its flow serves as its lullaby."

I want to linger in these moments of animal pleasures instead of the all-too-familiar precarity.

I want you to tell me of animal joys in Kallakurichi. What delights the buffalo, who Jaya Aunty loves? Do the dogs follow the sun to nap in its warmth? Tell me about the frogs on the full moon.

I can tell you about the pigeons on my windowsill, finding a safe place to perch on an A/C unit. Or the other birds who don't seem to mind bathing in the street puddles. I want to write to you about the animals at wastewater treatment plants. At all the ones I have visited, there are always cats—unofficial adoptees, communally cared for by staff. How do these interspecies relationships begin? I also think about the birds, the gulls, who love to feast on the water collection ponds. Who was the first bird that realized these new environments would be a source of food? How does this information get passed on?

We see raccoon eyes glowing in the dark at night, before they go to bed in the trees. There was that one time Mookie found two baby raccoons abandoned in a diaper box in the entrance to the park. We suspected one of them was dead. The other kept licking the sibling's static body, until she woke up. The two babies, born blind, immediately embraced. I want to imagine baby raccoon hugs in the Tamil literature.

But the Sangam poets also understood animal fear, grief, and suffering, millennia before modern Western science acknowledges the possibility of animals having feelings. "The poets have imagined the fish and other aquatic creatures in the tanks and backwaters as feeling afraid of the objects that look like their enemy," Varadarajan writes. "The shadow cast on the deep water by the rattan stem waving on the bund is mistaken by the valai fish for the fisherman's angle, as the poor creature has once had [his] own bitter experience of it." These poets are my forebearers in animal rights literature.

We learn how, when the male elephant slips and falls into a pit, his mate "breaks off big branches and throws them into the pit so as to improvise steps for the fallen animal to come up." Varadarajan describes the image of elephants encountering unfinished well pits covered with dry leaves, which they mistake for a snare placed by hunters, "and with great anger fill it up." The elephants become hunt saboteurs, like some activists we interviewed at *Satya*.

What can the Sangam poets teach us about animal grief?

"A female monkey is said to have lost [her] young one in a deep cleft and to mourn over [her] loss by crying aloud among [her] group," writes Varadarajan.

"On the death of [her] loving mate, another female monkey is imagined to hand over [her] young one to [her] kith and kin and perform sati by falling down from the top of a hill."

"A female crane is said to grieve for [her] mate caught in the net of some young boys. [She] stays with [her] young ones on the palmyra stem and cries aloud piteously without food."

---

As I write you, it is raining birds in India. Parched kites fall from the sky. Fledglings not ready for flight are fleeing their nests too early to escape the heat. I ask Sasipriya how to say *veterinary hospital* in Tamil. Kaalnadai maruthuvamanai—*the hospital for those who walk on their legs*. We, too, are animals.

From the local newspaper, I learn about wet-bulb temperatures, which consider heat, humidity, wind, and solar exposure to pinpoint the threshold of our survival, when sweat cannot evaporate and we cannot cool our bodies. I listen to a talk on care work and grief as a collective creative practice, as both memory and futurism.

I search the 3,675-page Intergovernmental Panel on Climate Change report to find 457 mentions of mental health, 23 instances of grief, and 3 of solastalgia. Philosopher Glenn Albrecht coined this term by combining *solace* (solas), *desolation* (desolare), and *suffering* (algia) to describe eco-grief and distress caused by environmental change. The Tamil Sangam poets communicated this precarity nearly two thousand years ago. Varadarajan reminds us that the deer "forgets the food of leaves and faints forlorn," in unbearable heat. The thirsty elephant, with a raised trunk, looks "at the sky for rainy clouds." She runs toward a mirage. Disappointed and defeated, she lies down like "a boat in a waterless river."

Some of these poets authored many Sangam verses, others, like the poet Vellādiyanār, have only a single verse attributed to them (or only one remains). I find it amazing that his indelible description, *"like a boat in a waterless river,"* once written on palm leaves, eaten by white ants and worms, survived two millennia, was reprinted, translated, commented upon, and landed on my bookshelf in a library that now sits upon a waterless reservoir.

≈≈≈

In search of parent rock, I look up the geology of Tamil Nadu. These stumblings and tumblings toward provenance lead me to a 2009 riverbed discovery of clusters of dinosaur eggs buried in the limestones of the Kallakurichi Formation. They belong to the long-necked, plant-eating, deprecatory-of-flesh-eating, sauropods.

I wonder if Thatha's divining rod would have sensed these football-sized eggs of our vegan ancestors? Year after year, placed layer on top of layer, the dinosaurs keep trying. Researchers deemed the unhatched eggs "infertile," a word that is often cruel and unimaginative. Try, bountiful elliptical orbs of knowledge that survived sixty-five million years absorbing and carrying the entire history of the earth.

≈≈≈

"If you look at the yogic view of the body there are actually five layers of this thing we call the body or the body mind," Sanjay says as he begins our second dowsing lesson.

"The one we are most familiar with is beautifully called in Sanskrit the Annamaya Kosha—*the body made of food.*"

With dowsing we want to invoke the next two layers—the Pranamaya Kosha, which is the layer of prana or qi, and the Manomaya Kosha, the layer of the mind. The fourth layer is the Vignanamaya Kosha, holding wisdom, and that final layer, the Anadamaya Kosha, contains bliss.

"One of the objects of yoga is to penetrate all the way through to an awareness of that bliss," Sanjay says. "When you do that, you become a blessing to the world."

Sanjay tells me water is a conduit for qi. "So, think of it, the water that's coursing through your veins in your blood has been here for billions of years. And it has flowed in the blood of so many beings on this planet. Dinosaurs, saber-toothed tigers, you name it, and here you are, you know you are. This miracle of you has that same blood flowing in your veins, and that blood has water that's within you that carries the energetic imprints of all the beings that it has occupied or possessed. And so, you are energetically this mélange from a million different things."

Sanjay's entry into this world of energy and healing was with his young son.

"This is the very first crystal I bought," he shows me, via video chat. "When I started exploring this with my son, he was going through his awakening, and I brought it home. Again, my son had it in his hands, and he was sitting in the chair kind of curled up, maybe eight [years old], and he went very silent as if he was somewhere else. And I got a little alarmed and so I said, 'What's going on?' He says, 'Dad, this is crazy. Inside this is a memory of a dinosaur and I can see in my eyes what the dinosaur saw.'"

"You know, it's because crystals, like water crystals, can also carry energy," Sanjay says. "So, the thoughts that we have are capable of being captured in the lattice structure of crystal. That's why my son was able to retrieve this dinosaur's memories from this thing, you know, if you believe that story."

I tell him dinosaurs have been on my mind. In my dual quests to find my ancestral and my vegan lineages, I learned about the original plant eaters—the vegan dinosaurs of Kallakurichi.

# THE TRUTH IS IN THE BODY

I OBSERVE OUR DOG, WHO FOLLOWS THE MOVING PATCH OF sunlight in the apartment from east to west, throughout the day. Another Tamil scholar, Xavier S. Thani Nayagam, teaches me how the Sangam poets paid equal attention to shade, noting the different kinds—thick and dense, or spotty and leafed—and how coverage changed daily and seasonally. Shade, he concludes, is akin to kindness.

Though their homelands were swallowed twice by the sea, the Sangam poets still viewed water as a gift. Sasipriya tells me that the Tamil word for *beauty*, ezlilili, is used by the poets to denote a rain cloud. "The cloud is the symbol of beneficence," writes Thani Nayagam. Rain clouds accumulate the wealth of oceans and redistribute them to the lands. A tender heart is described as a moist one, he says.

In Tamil, mei—the word for *body*—is the same as the word for *truth*. Sasipriya tells me it is what we call the consonants in the alphabet, too. I learn the vowels are uyir, *breath of life*.

The poets combine life with the body, life with truth.

On paper, the script looks like jalebis, waterfalls, and whirlpools.

I first learned to write the words for *birds*.

Their names sound like their songs.

*Crow.* காக்கா Kaakaa.

*Crane.* கொக்கு Kokku.

When I finish learning mei, *the truth in the body*, I can write *Kallakurichi*.

Will it be katai, *story*, kattukatai, *myth*, or carrittiram, *history*?

~~~

"Poitu varen," Sasipriya says when Tamil class is over. *I will go and come back.* "Because we believe words have power," she says, "we never say 'I am going,' without saying 'I am coming back.'"

Poitu varen.

# III

# UNGOVERNABLE BODIES

*For you, dear reader*

That this is your country, that this is your world, that this is your body, and you must find some way to live within all of it.

—TA-NEHISI COATES

And what a celebration when we realize that our survival need not make us into monsters.

—ALEXIS PAULINE GUMBS

# TRIVENI SANGAM

My gentle reader, we arrive at this confluence together, and it is you who I've been wanting to reach all along.

If you haven't realized it yet, this is a love story. I've been looking for the love between a person and an idea and the passion to live by this idea. I want to know that certainty to be called to do something, even when it's dangerous or risky. This is a story of love with a place that is both real and imagined. A place that was and can be. This is a story of finding language expansive enough to hold this great love for all beings and carry the accompanying grief.

We meet at this confluence of these three loves, at our own Triveni Sangam, like the mixing of Ganga and Yamuna with the invisible Saraswati.

I return to Howard Williams's *The Ethics of Diet: A Catena of Authorities Deprecatory of the Practice of Flesh-Eating*, and encounter a letter the ancient stoic philosopher Seneca wrote to his friend Lucilius about his preceptor Sotion's affection for the vegetarian teachings of Pythagoras. (This chain of thought is a catena in itself.)

> If these maxims [of the Pythagorean school] are true, then to abstain from the flesh of animals is to encourage and foster innocence; if ill-founded, at least they teach us frugality and simplicity of living. And what loss have you in losing your cruelty? I merely deprive you of the food of the lions and vultures.

This line—"And what loss have you in losing your cruelty?"—stops me every time. Can we, as a species, ever lose our cruelty?

I have been searching for this in the archive. The ways our ancestors (in blood and in thought) tried to lose cruelty. The variations of Kallakurichi under different names. The not-infernos amidst the infernos. But the archive has its own cruelties of erasure.

I write to you now, dear reader, as my attempt at creating a catena of kindness amongst a catalog of sorrows.

# DEAD ZONE

There's a growing dead zone in the Chesapeake Bay. Aquatic life is choked by chicken shit. One June day in 2008, I depart Brooklyn, in a silver Toyota Prius with a rounded front that resembles a baby elephant, and drive toward the heart of chicken country in Maryland to meet a visiting delegation of Indian egg producers.

Sixty-seven square inches. Less than a sheet of paper. That is the rhetoric animal activists use to describe the area in which an egg-laying hen spends her entire life, crammed in a cage, unable to spread her wings. In "The Chicken Issue," *Satya* published this spread, "Actual Size," that showed the size of an egg-laying hen's cage and the size of a rescued hen at Woodstock Farm Sanctuary, to scale. The healthy, liberated hen bulged out of the dimensions that confined her kin.

I had been growing curious about how factory farming was developing in India, the land of my ancestors, and the birthplace of so many ideologies around nonviolence. *Satya* magazine had closed the year before, but in our final year, we were tracking the global rise in the demand for animal bodies.

Seneca lamented this globalization of cruelty in his time, too. I found this in Howard Williams's catena:

> How long shall countless numbers of ships from every sea bring us provisions for the consumption of a single mouth? An ox is satisfied with the pasture of an acre or two; one wood suffices for several elephants. Man alone supports himself by the pillage of the whole earth and sea. What! Has Nature indeed given us so insatiable a stomach, while she has given us such insignificant bodies?

At *Satya* we were concerned with another trend—the humane-washing of animal agriculture. I found *humane* to be such a strange word, meant to describe the best of human behavior, applied to the worst of human actions. Two of my favorite issues of *Satya* are the ones seeking the truth in what was behind so-called humane, cage-free, free-range, happy meat.

*Satya's* closing was another great loss for me. I would reread the letters we published in our final issue as a reminder of our work. Here is one from Jeff Lydon, the executive director of Farm Sanctuary:

> I know you all have more stories to tell. All of us find comfort in the confidence that you'll keep fighting the good fight on another front on another day. [. . .]
>
> Mainstream publications leave a vacuum of authentic discourse. You filled it with integrity, intelligence, and a tough brand of grace. [. . .]
>
> Perhaps no group of progressives will miss your magazine more than we in the animal rights movement. *Satya* always had the guts to take seriously an issue that merited seriousness, no matter with how much contempt the issue was popularly perceived. A serious issue scorned—that pretty much sums up animal rights.

You've always been a class act. We thank you and we'll miss you.

I wanted to continue this good fight. I started researching the globalization of factory farming for an environmental action tank called Brighter Green.

I interviewed Chetana Mirle, an Indian American vegan activist who was working for an organization seeking to bring animal welfare reforms to India. She was arranging tours of several cage-free facilities in the United States for a small group of Indian producers. Animal farms in the United States were wary of visitors, so I was not allowed to join them on these tours. Chetana wanted the poultry producers to also see a farm animal sanctuary, where animals from factory farms were rescued and could live out their lives unharmed. She invited me to meet them at a sanctuary in Poplar Springs, Maryland. I decided to drive hundreds of miles to take Chetana up on her offer to meet the Indian egg producers.

When I pull into the driveway of the sanctuary, I find Chetana surrounded by three men. First there is Nitin, who previously worked for Skylark Hatcheries, one of the first and largest poultry operations in North India. She recruited him to come and work with her as a corporate marketing manager, to be a liaison to these large poultry companies. Surendra, director of Skylark Hatcheries and a self-described "poultry baron," joins them. He was part of that first wave of folks to bring these animal factories to India, but now he wanted to convert these operations to go cage-free. Jitender, an owner of a large egg-laying operation in Haryana, is also with them.

I am not sure what to expect meeting these barons, or what they thought of me. The only poultry barons I am familiar with are

Frank Purdue, Don Tyson, and the "Egg King of Petaluma" in John Steinbeck's *The Short Reign of Pippin IV*.

The Indian barons are soft-spoken, dressed in slacks and collared shirts, slightly formal for the farm sanctuary. Jitender is shy and has boyish features. Nitin assumes his role as liaison and translator.

We walk together to the chicken barn. The men look briefly at the birds, nonplussed. But when they see one that looks like a breed they have, there is a bit of interest. Surendra gently picks up a chicken and points to similarities in the feathers. He holds her briefly, and then puts her down. Jitender, who speaks only Hindi, is very quiet and just listens for the most part.

I was curious why India would look to the United States for animal welfare tips when India's constitution has language like this: "It shall be the fundamental duty of every citizen of India to protect and improve the Natural Environment including forests, lakes, rivers, and wildlife, and to have compassion for all living creatures."

But, like so many things, there is a gap between what is aspirational and the lived reality.

I began a series of inquiries as we stood around the chickens.

"Why do you want to go cage-free?" I ask.

The answer had nothing to do with animal welfare. "We will get premium price," Surendra says.

Nitin explains that 75 percent of all eggs are consumed in the big cities, where certain people can afford to pay a premium for these eggs. It's his job to educate them on the difference between caged and cage-free. Besides the cages, I am not sure there is much of a difference.

I ask Surendra how many egg-laying hens he had.

"Two lakhs," he says. Two hundred thousand. He has already piloted a cage-free system for about half of them, housing ten thousand birds in each shed. It's not the image of what consumers probably imagine cage-free to be. They aren't living like their wild jungle fowl ancestors in a rainforest.

I ask about forced molting. In the United States, egg operators deprive hens of food and light to trick their bodies into another egg-laying cycle.

"Skylark doesn't do forced molting, but majority of others do," Nitin says. "He thinks it is better to have a new flock." In a couple of years, India will ban this method outright.

I kept bringing up unpleasant practices.

"Do you debeak?" I ask.

"Yes. On the tenth day and the tenth week," he replies. To prevent these birds from pecking each other to death, their beaks are burned off with a soldering iron.

"What do you do with the male chicks?" I ask.

"The male chicks are separated," Surendra says.

"Then what happens to them?" I press. In the United States, chicks are sexed and sorted at birth, and the males are thrown into dumpsters or incinerators.

"They are sold in the market for a very low cost," Surendra replies. "The buyers take them to backyard areas." He then says something in Hindi to Nitin.

Nitin pauses and then translates, "If he goes for a big hatch, then we have to make them drown in the water."

I think about this language now. There is culpability (*we*), coercion (*have to make*), victims (*them*), and a typically-unspoken violence spoken (*drown in the water*).

Why do I focus on these hidden unspoken violences? If we see and speak about them, will that be enough to lose our cruelty? If that isn't enough, what is?

Our time at the sanctuary is about over, and we walk by the pig barn. Surendra stops talking in mid-sentence. He stares at an eight-hundred-pound sow sleeping in the hay. "I have never seen something like this before," he says. It is a rare sight for most of us to see pigs allowed to live and grow to this full size.

I say my goodbyes to the visiting delegation and thank Chetana for the invite. As I head out on my drive back to New York, my cell phone rings a few minutes later.

"We were feeling bad that you came all this way, and we didn't even have a cup of tea or lunch together," Nitin says, and invites me to meet them at their hotel in DC. I agree, and Chetana and I are invited up to their room for chai and snacks. The poultry barons complain about the lack of hospitality in America. They have been visiting all these egg facilities and no one has even offered them a cup of tea.

"In India," Chetana says to me, "you can go to a slaughterhouse, and you will be offered tea and snacks, whether you want them or not."

Eating in the United States had been hard for them. Jitender, it turns out, is a vegetarian. Nitin is a Jain and had previously eaten meat, but is now trying to go back to vegetarian. Surendra eats most things but avoids beef. They've been subsisting on nuts and dried fruit. It isn't what I picture the life of the traveling poultry baron to be. It is more similar to the life of the traveling vegan.

They bought strawberries from the Amish farms in Pennsylvania and offer them to us. Nitin makes chai without milk for Chetana and me.

We sit in a circle, talking about labeling schemes and consumer fraud. Chetana says people just put whatever labels they want on eggs. She saw a guy selling "Bird Flu Free" eggs. There is no regulation or certification for that. Nitin talks about one guy selling "Natural" eggs, meaning they came from hens and not machines. "People don't know where eggs are produced. People think eggs come from the ground. They don't know how hens lay eggs," Nitin says.

There is also the belief that brown eggs are "desi" eggs and white eggs are foreign eggs, and people pay more for their patriotic desi eggs. Everyone laughs.

Chetana suggests we go out for lunch. The men are still wary of American fare. "We'll get Indian food," she says, hoping to allay their concerns. Jitender, who has not said much of anything the whole morning, starts to smile. His face lights up.

We share an Indian meal together and I begin my journey back home to Wan and Moo Cow.

When I arrive back in New York that night, I check my voicemail. There is another message from Nitin. The poultry barons were checking to make sure I got home safely and extended an invite to see their farms the next time I visit India.

# ANOTHER DEAD ZONE

When Wan and I land at the airport in Bangalore, a man wearing a white mask aims a gun at our heads. Temperature scans. Swine flu. It is 2009, and just a few months prior, an outbreak of H1N1 broke out in Mexico. A cluster of infections surfaced in New York City, and the World Health Organization declared a public health emergency. At the airport, the only line of defense screening potential vectors from 1.2 billion people is a short paper form inquiring whether we have any symptoms and this man with the temperature gun.

It is the summer after our wedding. Before we arrived in Bangalore, we spent two weeks in Korea. In their retirement, Wan's parents started a small organic bokbunja (*blackberry*) farm in the foothills of the Taebaek Mountains, in the eastern province of Gangwon-do. There, we picked and enjoyed—with purple-stained fingertips and lips—the fruits of such labor. In addition to bokbunja, my Korean vocabulary is comprised largely of vegetables: baechu, oi, gochu, pah, goguma, ogsusu—*cabbage, cucumbers, peppers, scallions, sweet potatoes, corn.* The fabulously named sinseon i meongneun yakcho (*herbs that mountain wizards eat*) also grew there. Various pepper and bean sauces marinated in the sun in jangdok, traditional Korean ceramic jars. There was no reliance on chemical fertilizers or pesticides. Their farming was rooted in the belief that if you take care of the land, the land will take care of you. They work hard, take rest breaks, and enjoy drinking blackberry juice and sharing what they grow with their neighbors. Eomeoni, my mother-in-law, says she feels like she lives in a poem. My father-in-law Abeoji says he feels like he lives in a novel.

Our time in India would be different. I called up the poultry barons to visit their facilities. Their farms are not unlike the one in Mexico where this recent swine flu outbreak initiated. *Farm* is not the right word. There are other names: Animal Factories, Industrialized Animal Production Facilities, and Concentrated Animal Feeding Operations. These are not places where you pluck blackberries and make wine, or where mountain wizards consume herbs. You could write poetry or novels about these places, but you would never delight in living in them.

Before we arrive in Delhi to start this research, Wan and I spend time in Bangalore. We ride in traffic alongside two-wheelers and auto-rickshaws. Stray dogs run parallel to the car. We watch cows foraging in the street on household litter wrapped in plastic bags. A young boy, maybe six or seven years old, rides a purple-and-pink bicycle next to us, wearing a thin red towel, like the kind my mother uses to dry her hair, wrapped around his waist. Behind him are the wheels and carts of a carriage that has been decorated for some occasion. A dog is sleeping behind one wheel. Two goats are tied up behind the other. A horse is standing in the backdrop.

I think about how animals are everywhere in this cityscape, yet also oddly rendered invisible. I want to know their stories, how they see and navigate the city and make sense of this changing country.

One evening my family and I go out to see the latest Harry Potter movie at one of Bangalore's newest shopping malls. When we arrive at the food court, Thripura Aunty turns to me and says, "This looks like America, no?"

*Globalization* is perhaps the catchall phrase that captures all the forces responsible for that particular moment. But it isn't just about watching Harry Potter in an Indian shopping mall. Globalization is also about that thermometer gun at the airport—how a virus in a pig in Mexico could mutate and jump species and end up jet-setting across the world, like we do. Animal killing factories are being exported to India along with Harry Potter and Happy Meals. I came to understand this.

---

A golden Ganesha pendant dangles from the bell hanging from the rearview mirror of Nitin's car. As he swerves in and out of New Delhi traffic, the jingles interject and punctuate our conversation.

I do not know if Nitin and the poultry barons expected me to really come, but they make good on the promise they made a year ago. Nitin picks us up from the airport.

"Guests are like gods," Nitin says. "In India, it is an honor to have guests."

During the car ride from the airport, Nitin's mobile phone rings, and he switches from English to Hindi. My Hindi vocabulary is like my Korean. The words I understand are food items. Dhal. Roti. Chawal.

"They are getting lunch ready and wanted to know what we wanted," Nitin translates after hanging up the phone. Ganesha's bells continue to jingle as we drive. I reach for the dangling elephant-headed pendant and stop the jingling.

"We are going to have lunch there?" I ask. "At the chicken slaughterhouse?"

"Don't worry. They know you both are vegan," Nitin says.

I remember the barons' trip to the United States and how they had complained about the lack of hospitality. Chetana wasn't exaggerating when she talked about the slaughterhouse etiquette in India.

As Nitin drives on Delhi highways, he points out all the construction. Flyovers in progress. *Inconvenience is regretted.* A landfill on the outskirts of the city. Mega shopping malls. The rise in consumption of animal products parallels the rise in urban incomes. But Indians still don't consume at the same rate as Americans, who eat thirty-five times more meat per capita. Eating meat in India now represents (or maybe has always represented) multiple things: a status of wealth or sustenance for the poor; tradition to some and a rejection of tradition to others; an act of violence or a reaction to violence; a marker of religion or caste or the shedding of those markers. But for the animals, it means the same thing it always has.

What I will learn during this trip is how even those who abstain from eating animals can still profit from them and be responsible for their deaths.

I want to see the animals and conduct interviews, and Wan wants to take photographs. We think of Jane Goodall and her wildlife photographer husband Hugo van Lawick. Perhaps we can be this kind of duo for chickens and cows.

I discussed with Nitin visiting chicken farms, dairies, and piggeries. With some hesitancy, I added slaughterhouse to the list of places to visit. I thought we could do that last, if I mustered the courage to enter. But Nitin called the day before we arrived in Delhi, saying the only way we would be able to get in the slaughterhouse is if we went first thing straight from the airport.

"Monday, the manager won't be in to answer any interview questions," he told me. "And Tuesdays they don't slaughter because it's Hanuman Day."

"Every Tuesday?" I asked.

"Yes. No cutting on Tuesdays. No barbers or butchers are open," Nitin said. We planned to cover 1,400 kilometers within five days and couldn't wait until Wednesday to start that journey. "It's best we go on Sunday and hit the road on Monday," Nitin said.

There were many things in this conversation I didn't understand (and still don't). What does Tuesday have to do with the monkey-faced Hanuman? Why does Hanuman object to slaughter (or haircuts) on Tuesday only, and not every day? But I had no time to discuss all this in that phone call. I just gave Nitin an answer.

"Yes. We can go," I said. "We'll see you tomorrow."

So here we are now in Nitin's car, and I am still trying to wrap my head around the news that a vegan lunch is being prepared for us at a slaughterhouse.

"The owner of the slaughter plant is a vegetarian, so he has a separate kitchen," Nitin reassures us. I don't know what to do with that information.

The slaughter usually takes place in the morning. It will take another hour and a half to get there. Nitin is rushing to make sure we can see the process.

"They are almost done with the slaughter, but they are holding two hundred of them for us to watch," Nitin tells me.

I take hold of the jingling Ganesha once more, as if the silence would help me understand. I feel sick. The plan was to observe a typical day's work, but to have two hundred held for my benefit seems wrong, as if it makes me complicit. *Could I really just stand there and watch?*

The reasons not to go into a slaughterhouse are clear, and people usually fall into two camps: those who don't really want to see what goes on in there (even though they already know), and those who know, and don't need to see.

I think it's important to bear witness. *If the eye sees, the hand must fix.*

But I also wonder, do we need to see violence to fix it? I have spent the better part of my adult life learning about the abuses animals endure at the hands of humans. It felt like a responsibility to know. In the early years, I absorbed each bit of knowledge, and it fueled me. But at some point, the same truths that once sustained us, can harm us, if we are not careful.

How do we see the truth and protect our hearts?

Nitin pulls into a gravel parking area outside a gated warehouse-like compound. Women in bright-colored sarees sit on the grass eating their lunch, draping their sarees over their head, shielding themselves from the afternoon's rays. At the entrance gate, we dip our shoes into a footbath—a tray of water. The manager of this processing plant, Mr. Harinder, greets us on arrival.

In Mr. Harinder's office, a giant poster board explains their operations "From Farm to Fork." Skylark is implementing "vertical integration" and controls every aspect of the chicken industry. They have the first and largest automatic feed batch plant in Haryana with soy and corn from Rajasthan, Bihar, and the southern states. They use a particular genetic stock of birds imported from Europe and breed them in their hatcheries in India. The chickens are raised in contract farms and then transported here to this very modern abattoir. They leave this plant—fresh or frozen—wrapped in plastic with the label "Nutrich Hygienic Chicken" and end up on the shelves of a national grocery chain, where they are sold to a growing affluent elite who can afford a premium for "hygienic" chicken.

When I read "vertical integration," my mind thinks of Bhanu Kapil's *The Vertical Interrogation of Strangers*. Who is responsible for the suffering of these chickens' mothers?

We take a seat in Mr. Harinder's office. I can't find my pen, and Mr. Harinder gives me one with the Nutrich Chicken name on it.

I open my notepad and start asking questions about the plant. Mr. Harinder brushes me off for a second and phones his assistant to bring us drinks. At first, I think he is evading my questions, but he is just being a good host.

"Every place we go, we will spend at least forty-five minutes with such protocol," Nitin says.

A round of water arrives. Black tea comes next.

Mr. Harinder speaks about this plant with pride, explaining how it differs from Ghazipur Market, the largest live poultry market on the outskirts of Delhi. The majority of the city's chicken meat still comes from Ghazipur. Every day, one hundred thousand birds are killed there in the heat and open air. Mr. Harinder used to work at Ghazipur, selling chickens wholesale.

"Blood and guts everywhere," he says. "Manual slaughter is very cruel." He shrugs as if trying to shake the memory.

Nitin said he couldn't sleep for three days after visiting Ghazipur Market. Chickens are being annihilated in the same building as we all are sipping tea. How was this slaughterhouse any different from the blood and guts of the live market, which haunted their dreams?

"Come, let's go." Mr. Harinder motions toward the door. An assistant enters, bringing us our new outfits. Wan and I each put on a smock, shower cap, face mask, and shoe covers. We look like doctors prepping for surgery. Nitin decides not to join us. He waits in the office.

Wan is going to take still photos, and I want to shoot mini video clips. But I am clumsy—too bewildered to hold the camera steady while taking notes with my Nutrich pen and asking questions. Wan takes the video camera away from me.

"Just go ask questions," he says. "I'll do both video and photos."

Wan knows me. I am the one who holds my breath and bolts past the meat section in the supermarket. Now I have to keep my eyes open and take it all in.

Our tour appears to be in reverse. We first see the lab where they do salmonella and E. coli checks on the final product. When we enter the processing floor, the buzzing is overwhelming. Compressors. Chillers. Refrigeration. We walk into the freezer and see rows and rows of frozen headless and featherless bodies. These bodies enter a conveyor belt where they are chopped, packaged, and labeled in assembly-line form.

Many of the workers at this plant are migrants from Nepal. Some live on-site, in quarters provided by the processing plant, and return to Nepal one month out of the year. About half of the hundred workers in the plant are local women who trim the fat off the birds.

We walk to another room. Mr. Harinder talks us through the operations. It is loud and difficult to hear, and because he has a face mask on, I can't read his lips. I write down the names of the processes.

*Defeathering.* A mound of feathers covers the floor. After the birds' throats are slit, they are dropped in a scalding tank of hot water, to remove their feathers.

"What do you do with the feathers?" I ask.

"Send them to the rendering plant," Mr. Harinder says. "Sometimes they go into bird feed." Chickens will be fed the feathers of their kin.

*Evisceration.* This is where the internal organs are removed. Blood and guts still everywhere.

When we finally arrive at the live transport–unloading dock at the slaughterhouse, it is empty. They didn't wait for us. They had already slaughtered all the birds, even the last two hundred.

Mr. Harinder describes to us what the process would have been. The birds would be stunned, cut in the jugular, and sent to the scalding tank. Each hour, 1,350 birds are stunned and killed. It is the most gruesome aspect of the operation, but we do not witness it. He gives me a CD to watch. When I return home to NY, I put it in my computer, but it is blank.

What we see instead is a mound of excrement by the unloading dock. In their last moments of fear, this is the last shit these birds take before their necks are slit and their bodies boiled, defeathered, eviscerated, chopped, frozen, and packaged as nutritious, rich, and hygienic.

Two tiny birds hover in this area. They are too small and aren't deemed profitable for slaughter, so were just let loose. They, like us, were spared.

"We can go back in now. Lunch is ready," Mr. Harinder said.

# BAS BAS BAS

"Okay, time to move," Nitin says. He wants to make it to Kurukshetra before nightfall. "It is a very famous place," Nitin explains. "It is the battleground in the *Mahabharata*."

I confess I haven't fully read the *Mahabharata*. It's hard for me to get into these epic battles, but Nitin tells me Kurukshetra is where Lord Krishna delivered the Bhagavad Gita to Arjuna. King Ashoka also deemed Kurukshetra a center of learning for people around the world. Ashoka, I know.

Years later I will pick up *Until the Lions: Echoes of the Mahabharata*, Karthika Naïr's radical retelling of the epic.

In Kurukshetra
the Earth swathes her face in blood,
Death begins to dance.

This is the Kurukshetra I will come to know. The Kurukshetra of the hens. I don't know if they ascend to heaven, as those slain in Kurukshetra were rumored to do, but I do know they have already been through hell.

"Pummy will be hosting our dinner for tonight," Nitin says.

He is the soft-spoken, quiet poultry baron named Jitender I met the previous year in Maryland. We drive to another hotel to share another meal together. In Nitin's car, the Ganesha bells jingle again as we ride along National Highway 1 in Haryana. "Water. Water," Nitin points out. Nitin knows I am interested in water use, waste, and pollution. Periodically, he shouts "water" while driving.

An egg factory is adjacent to the highway, and water runoff extends roughly a kilometer in each direction. Nitin tells me this place is filled with rats.

Night is settling in quick, and we have to wait to visit this place in broad daylight with the rest of the facilities on our egg tour the following day.

"You'll see more tomorrow," Nitin says.

We check into the Saffron Hotel in Kurukshetra. Wan and I wash up and then head to Nitin's room. We've only eaten a couple of bananas all day. We order drinks and pakoras while we wait for the poultry barons to join us.

Jitender arrives with his friend Som Parkash. Nitin and Jitender hug. Jitender is laughing and smiling. He is so much more comfortable in this hotel room than the one in Maryland. He brings us juice to share. Unsure of what we would like he came with three varieties: mango, guava, and mixed fruit.

"It's so good to see you again," I say.

"Call him Pummy," Nitin says. "P is for *Pummy* and for *Poultry*."

We order another round of drinks and snacks and try to cool off from the sweltering day. The A/C is on in the hotel room, but the power cuts off every ten minutes or so. When the snacks arrive, Pummy starts loading my plate with chickpeas.

"Bas, bas, bas," I say. *Enough*. Laughter erupts. My limited Hindi is a big hit.

"If you have any queries, now is a good time," Nitin says. "Tomorrow, they will take you to their farms, but it's better to ask your questions now."

I take my notepad and my new Nutrich pen out of my bag. I draw a line down the middle of the paper. One side will be Pummy's answers; the other side for Som Parkash.

"How many birds do you have?" I ask, and Nitin translates.

Twenty thousand for Pummy; thirty-five thousand for Som Parkash.

"When did you start?"

1990; 1987.

"How many did you have when you started?"

Nine thousand; thirteen thousand.

"Why did you start this business?"

Som Parkash answers in Hindi and Nitin translates. "Because in 1987 and 1990 the poultry graph was going up. Feed cost was so low." The Indian egg industry has an 8–10 percent growth in consumption every year, and a 25 percent growth in production, they tell me. Pummy took loans from his brother to start the business.

"Who eats your eggs?" I ask.

They are sold in Delhi and Uttar Pradesh, in open markets.

I ask them to walk me through the life of an egg-laying chicken. They both receive day-old chicks who have traveled seven hours from hatchery to farm. The chicks brood for four months. After five months, they start laying eggs. Their hens lay eggs once a day, but as they get older, it takes more time. When she is no longer as productive, they get rid of her.

"How old is that?"

About eighteen months.

"What happens to her then?"

The middleman takes her away to be slaughtered.

During this trip, I hear lots of talk about the elusive and essential middleman. The middleman who does the job of disconnecting the businessmen from the realities of their business.

"Who eats this meat?" I ask.

The military, they tell me. The meat of spent egg-laying hens is cheaper than broiler meat.

Both Pummy and Som Parkash use battery cages to house their hens. Nitin is trying to get these guys to go "cage-free."

Som Parkash says he started with cage-free. Those were "deep litter" operations, but it required egg collection six times a day. With caged hens, collection is only twice a day. Labor is getting too expensive.

"Where is your labor from?"

Uttar Pradesh; Uttar Pradesh and local Haryana.

"Now the labor issues are rising a lot," Nitin says. "We are facing a labor problem in India. Cage-free needs a lot of labor. But going for automation needs a lot of money. They don't have such money to go for full automation."

Now other industries are booming, so why would people want to work in poultry farms for twelve to sixteen hours for such a low wage, when they could get an industry job for eight hours for more money? Nitin explains.

"Do you have anything else? Shall we go to dinner?" Nitin asks.

I have one more question, but am unsure how to ask it, so I wait.

We go to a nearby restaurant called Hot Millions. Bryan Adams ballads play in the background. We order dishes to share. Wan and I choose off of the Indo-Chinese menu. Hakka noodles. Fried rice. Gobi manchurian. Out of habit, I blurt out the words to the waiter, "No eggs." I look across the table to see if there is a reaction from the egg barons, but it turns out they don't eat eggs either.

"They are from Arya Samaj community, which has a big respect for animals," Nitin says. "Their community is very strict. They don't eat eggs. They don't drink alcohol."

Now is the time to ask the question I didn't know how to.

"You don't even eat eggs, yet you are in this business?" I say.

Nitin starts to answer for them. We had a similar conversation in the car, when talking about the vegetarian owner of the slaughterhouse. These things didn't surprise Nitin. "Even my sister—she has never touched an egg in her life—is a director of Skylark Hatcheries. Many people are like this."

"But I want to know why *they* started this business," I say. Pummy is quiet during this time. How does he reconcile these two lives? Som Parkash responds straightforwardly in Hindi.

"From the guidance from a good friend from Punjab, he was told to invest in poultry. That it would be profitable. They don't eat eggs. They don't let eggs into their own home," Nitin translates. "The poultry business is in cash dealings. No credit. No government regulation. No income tax. It is a good business. One person in the community started it, then helped the others join."

When he first started, Som Parkash made a good fortune in the first year alone. Now things are changing. Over the past few years, the egg industry is facing more problems. With the rising cost of feed, shortage of labor, and increase in threats of disease and public health scares around bird flu, profits are dwindling and some facilities are closing. They are now at a crossroads: get big or get out. I am reminded of Steinbeck's Egg King of Petaluma:

*Chickens are luxury until you have fifty thousand of them. With that number, you may break even. With one hundred thousand you can show a small profit. Over half a million, you begin to get some place.*

The next morning, we wake up and check out of the hotel and head to Pummy's house for breakfast. Nitin prays for a minute in his car, before swinging into reverse, and the Ganesha bells start ringing again.

We enter a neighborhood of estates. Pummy and some of the other poultry barons in the community live here. There is a big Om sign in front of the gate of Pummy's house. Pummy greets us at the door and escorts us to the living room. We meet his wife Kavitha, who is making breakfast for us—aloo paratha, dhal, mixed pickles, and rice. We leave their meatless, eggless home and climb into Pummy's car and begin our journey around the egg factories of Kurukshetra.

We tie plastic bags around our feet and enter the gate to Pummy's poultry farm. Pummy has a two-story brick building to house the birds. Another building is under construction, and workers pass mud in buckets to the second story for bricklaying.

Chicken waste piles up on-site and is sold to farmers for manure, but it is also carried away with rainwater runoff onto adjacent rice fields and drainage channels. Nitin jokes that the chicken shit ended up in the rice we had for breakfast and will have for lunch.

At first, I take overview shots with a video camera. Trying to get a sense of scale and the big picture. Three rows of cages, each with three tiers of birds, span the length of the building. Wan zooms in close, to focus on a few birds at a time. In each tiny frame, their stories are told. *The truth is in their bodies.*

We notice the bare spots on parts of their bodies where their feathers have fallen off, some with open sores. Three to four are packed into a cage. In some cages, one bird is on top of the other. Mortality on the farm is two to three hens per day, about one hundred per month.

Thoughtlessly, I decide to walk along one row with a video camera and pan across the length of the building, capturing the birds in each cage of one tier. I get about halfway down the row when the screams of the chickens overwhelm me. I am a new person in their place. One by one, they all start fluttering wings that cannot fully spread. Particles of feathers, dust, and what must be chicken excrement land on my clothes and face. The screams intensify. I panic and freeze, unsure of what to do. Do I continue down the row until the end? Then what? I would have to make my way back somehow.

Pummy notices my predicament and walks toward me. He gestures to me to walk slowly and relax, and the birds calm down in his presence. As for me, I am drenched with sweat and caked with the fear of hens.

"Perhaps they thought you were coming to cut them," Nitin says.

Someone else would do that soon. The birds on this floor are nineteen months old and the middleman will take them away the following week.

Our next stop is an infamous egg farm whose name Nitin asked me to keep anonymous. This is the place with the rats and the pollution runoff adjacent to the highway. Nitin calls it a "three-story farm."

It is mid-morning by the time we arrive. When we enter the feed storage area, we see the rats. I find them scurrying along planks on the ceiling and on the floor. Rat carcasses lay in the doorway and in various corners of the building.

A young man from Uttar Pradesh collects eggs. He is wearing flip-flops with no face mask or gloves. Eighteen years old, he has been working here for two months.

There are fifty thousand birds at this facility. The smell of the excrement is overpowering. Having visited the first and second floor, we ascend to the third floor, the worst of the three. The aisle between the rows of birds is filled with wet excrement. On this floor, the egg collector wears boots, which are covered up to his ankles with chicken poop.

The owner of this farm does not come out to meet us or offer snacks. Nitin thinks he is too embarrassed. This particular egg baron is a Jain. He adopts a home life that minimizes harm to all living beings, and a business where suffering is palpable from all the senses.

We hurry back to the car. I have a thin, red hair towel, to wipe the sweat, dirt, and chicken shit off. Nitin comments that it looks like the holy cloth carried by religious pilgrims. As we continue this pilgrimage of a different sort in Kurukshetra, I take out my hand sanitizer, and Nitin, Wan, and Pummy cup their hands waiting for me to pour.

Next, we visit Som Parkash's farm. He is happy to see us after our dinner together the night before. He invites us into his office for snacks and to drink the leftover juice that Pummy brought. He has a bowl of glass and marble eggs decorating his desk.

"Feng shui," he says.

Pummy and Som Parkash compare water usage. While Som Parkash has more birds, he uses less water because he feeds the birds with a nipple-drip system, where Pummy has a constant flow running from the tap into their water channel feeders.

Som Parkash tells us he used to be a reporter for a Punjab newspaper in the late 1980s, before he arrived in Haryana and invested in the poultry business. I am relieved to sit for a moment in his office before going back into another battery egg facility. I've seen enough, I think. *Bas. Bas. Bas.* But every time I have this thought, I discover something new.

Like, in that next farm we visit, we see a few uncaged hens roaming or sitting in the aisles between the caged birds. These are sick or lame birds who have been given some space in an attempt to heal them. Nitin picks one up. She is dead. Another can't walk and has an open sore that is covered with dirt and bugs. That farm was hit with Gumboro disease recently, even though the birds were vaccinated against it twice. Twenty percent of the forty-five thousand birds died in one week.

Suresh Garg, a vegetarian Agrawal Jain who also doesn't eat eggs, runs this farm. He grew up in Punjab. But after the Khalistan Riots in the late 1980s, he moved to Haryana and entered the poultry business. His story is similar to the other barons. Relatives encouraged him to join the poultry business. No sales tax. No income tax. Cheap inputs.

In the last facility we visit, Nitin comments on how meticulously the records are kept on this farm: the number of hens, number of eggs, mortality, etcetera. I imagine a fancy spreadsheet or database, but the records are not electronic, and use none of the advanced computing skills modern India is revolutionizing. Instead, the data is recorded in a child's school notebook. The cover of the book has a picture of my favorite goddess of books,

Saraswati, and the quote, "We serve the nation through education." In the box for Name/Subject, the word *egg* is written.

Om Shanti (*Peace*) Farm, as it is called, was formerly a rice shelling operation and switched to poultry ten years prior. This egg factory is the largest one we have seen, with six sheds each two stories high, housing one hundred and fifty thousand birds. They previously used a water channel system to feed the birds, but water and waste was running off to neighboring properties, attracting houseflies. The village residents protested, breaking the windows of the farm's office. Soon after, a nipple system was put in place.

Women in bright orange and green sarees collect the eggs from the cages. A young girl dressed in a school uniform with her hair in two long braids grabs an empty carton and runs along the corridor of cages taking eggs from the hens, playfully imitating the older women. Young girls carrying babies on their hips stare at us. Four or five kids play in between the buildings housing the hens. They stand in a circle and laugh. Among the feathers, dust, and ammonia, they create a playground.

Can these hens, trapped in the inferno created by others, find a moment, however briefly, to orient themselves toward joy?

There is a hike in grain prices and the feed costs scare the industry. Changes in weather are part of that uncertainty. "Climatic changes will affect a lot of breeds," Surendra told me the year before. "The time of summer months is increasing. Birds die due to heat stroke. In India, the last two-to-three years, we are not getting proper rains. Due to that, we are facing huge problem of the grains."

It is getting more costly to feed the birds. It feels like a system on the verge of catastrophe. The poultry industry is not environmentally regulated in India, either, and pollution has become a big problem. "There is no awareness of the pollution in India," Nitin tells me.

The Yamuna River would become their Chesapeake Bay.

"Iyer, you're sketchy," Wan says to me after our journey to the land of battery egg cages. It is a few days before our one-year wedding anniversary, and I have been dragging my husband around shit and death factories. He wants to go see the Taj Mahal when this is all over.

"People come to India from all over the world to see the Taj. We come to India to see chicken farms. *Anywhere* else is better than a chicken farm," he says.

*Bas. Bas. Bas.*

We are in a hotel room, handwashing our clothes from the day. This is part of the ritual we establish for the week.

Sometimes we recount what we have seen. Sometimes we are silent, in the comfort that the other knows what happened that day, without having to relive it.

We transfer our pictures from the camera to our computers and backup drives, and charge our batteries overnight. I type up the notes I jotted down.

This is how we preserve the memories of the day, before we scrub our bodies at night trying to erase them.

---

In the town of Ganaur in the state of Haryana, a number of egg-laying facilities are abandoned. Fluctuation in prices and outbreaks of avian flu drive some barons out of the business. Their two-story brick structures now lay empty and silent.

One owner, however, has a different idea. Anol Chadha converted his twenty-five-thousand-hen operation into a piggery with eight hundred pigs. We sit in his office with a tray of mango juice, water, and Mountain Dew in front of us. A Ganesha calendar for a handloom retail company hangs on the wall behind him.

Chadha said that, while Muslims abstain from pork, Christians and some Hindus eat his pig meat. There was also a growing number of foreign nationals and students in New Delhi who consume pork. I was also told that pork (and beef) is often mixed with or labeled as mutton and sold in the market to unknowing customers.

Outside his office, piglets, some only two to four days old—tiny, pink bundles—latch on to their mothers for sustenance in their pens. When they reach 90–100 kilograms, approximately at the age of nine to ten months, they would be slaughtered.

There are no pork processing plants in that area. The modern abattoirs that kill chickens often export to the Middle East, and do not allow pigs to also be killed there.

Instead, the pigs are manually sawed: Several people hold the pig down and tie her legs with rope. Her head will be cut off, which is a slow and painful process. Then her stomach will be opened, her intestines removed, and her body will be cleaned, chopped, and sold by the kilogram.

But Chadha only sees the tiny pink lives, and not this end of the process, where the middleman takes them.

We happen to visit this piggery on the same day swine flu claims its first human fatality in India. Could a temperature gun alone protect the world from the health and ecological risks of these systems that confine and control animal bodies?

I wonder if the barons really want this as their future. One of the last things I ask Pummy is whether he hopes his children take over this business. Pummy wants his kids to go to college and do something else. He clasps his hands together and raises them in the air.

"Do you know what that means?" Nitin asks. "It means he is fed up."

*Bas bas bas.*

# BROOD

WHEN A HEN IS BROODY, SHE SITS ON HER EGGS, PREPARing for them to hatch. Her young are also called a *brood*. When a person is broody, they are thoughtful but unhappy. *To brood*, as one definition suggests, is to think deeply about something that makes one unhappy. I have been brooding about chickens in India. About their screams, the loss of their plume, the sores on their feet. About living in their own shit.

I've been brooding about the vegetarian meal at the slaughterhouse. About the egg barons not eating eggs. I mentioned before how satyagraha could be weaponized. Ahimsa can be, too, when it operates in this sort of NIMBYism—*Not in my backyard*. The barons can keep the eggs out of their own home, out of their own mouths. They may not touch eggs or chickens, but are still responsible for what is done to the birds by other hands.

I'm less interested in pointing out the inconsistencies in these few men and how they compartmentalize their lives. It's about all of us. How do we bear it? In most cases we don't. It's too much for any one soul to take. Most ignore, or as Calvino says, they become such a part of the inferno that they can no longer see it. How do we not become the inferno here? Why do all these concepts—*a*-himsa, *a*-shoka, *non*violence, *not*-inferno—position us in relation to harm? Is there some other way? Can we go where kindness and regard are centered? How do we go from here to there?

# IN SEARCH OF SACRED COWS

There is an Italo Calvino story about stubborn cats in his book, *Marcovaldo*. Calvino's narrator follows a stray cat around for a day and ends up finding this hidden cat community. I wanted to do something like this in India, not with cats, but with the street cows. I wanted to wander these invisible animal cities, that are both hypervisible and hidden at the same time. In India, though, I encounter cows at different parts of their lives in different places. Not the same cow or city or street.

After learning about my interest in cows, my mom's sister, Thripura Aunty, would point to every cow we see on the streets of Bangalore and say, "cow," like Nitin would shout "water." My cousin took me to a local dairy—Nandini: Hygienic Milk and Milk Products for Marriages and All Functions. She spoke with the owner and told him, "My cousin came from America, and she would like to see your cows." The cows were not there when we arrived and were out rummaging through garbage on the streets.

On a different day, in a different city, I find out what could happen to some of these street cows.

A truckload of cows arrives at the Gopal Gausadan shortly after we do, on a sweltering August day. This government-funded gaushala, or cow shelter, houses roughly twenty-five hundred cows. Eight soon-to-be residents are in this truck. Earlier that morning, each was on a Delhi street when a government-hired urban cowboy wrangled her into a truck and inserted a microchip down her esophagus. The Municipal Corporation of Delhi (MCD) orchestrates these roundups and transports the cows to shelters like this one, to live out the remainder of their lives. A few hundred arrive every month.

Mr. Mahavir, who opened this gaushala in 1994, greets us at the entrance to his cow house, draped in a white loongi. It is early morning and almost a hundred degrees. Stainless steel tumblers of water are presented to us. We know this water is scarce, as we are in near-drought conditions. On our drive there with Nitin, we spotted women lining up with metal jugs on their heads to get their ration of water from government supply trucks.

Remember these lines from the *Tirukkural*.

*No being can be without water—nothing can flow*
*For anyone without rain.*

A young man at the gaushala climbs into the truck and scans the belly of each of the eight cows to record her microchip number. Then, the back gate of the truck opens, and one by one, each cow leaps into the large shed where the other cows are resting.

We next visit an outdoor pen where all the bulls are contained. They were the unwanted males of the dairy industry who escaped starvation and slaughter. They come rushing to us, like giant puppies. Their large bodies are packed in together and it's like one big mass of bulls.

Wan and I participate in snack time. We hold large, brown jaggery cubes over the fence and hand-feed them. There are dozens of them corralled into this one area. They lick our hands gently and devour their treats, like our beloved pit bull Moo Cow would. They are intent on feeding, but so careful not to harm us. They press their faces to our hands and touch their noses to our palms.

We let the bulls lick the remaining sugar off our fingers.

When Gopal Gausadan first opened in 1994, the government gave it eighty-five acres of land for the cows to graze. In 2006, sixty-five acres were taken back and allocated to a government reforestation project, which was mitigating impacts from the construction of Delhi's metro. *Inconvenience is regretted*. With the cow shelter's area so vastly reduced, the cows no longer grazed freely.

To stay in operation and purchase feed for the animals, the gaushala pursues additional revenue schemes.

I was surprised to learn they also run a small dairy, considering gaushalas are supposed to be a refuge for the discards of the dairy industry. About forty cows are kept in milk production. Wan, Nitin, and I are brought to the canopied area where the dairy cows are corralled. Their babies are kept in a separate area. One of the staff workers catches a calf in an attempt to impress us. The baby is emaciated, with protruding ribs. He is only four days old.

"Where is his mother?" I ask.

"Would you like to take a picture of the two?" the staffer asks, again attempting to please the visiting guest. Guests are like gods, as Nitin would say.

He brings the mother to the calf. The instant the mother walks into the pen, the calf starts suckling. It looks magnetic—how quickly their bodies come together.

"Why do you keep them separated?" I ask.

"If we kept them together, then he would be suckling all day," he says, as if her milk was not for her calf.

I remember reading an article in Gandhi's newspaper *Harijan*. The reporter, Satish Chandra Das Gupta, visited a flaying yard in Bombay, where dead animal bodies were processed. Among the heap of carcasses, he observed many calves. "This preponderance of calves amongst the dead cattle shows that these were allowed to die of neglect and starvation. The buffalo are milked to exhaustion to supply the city with milk, while only a little portion left with the mothers for the calves would have saved the lives of so many calves daily," Das Gupta wrote in 1934.

Many decades have passed. Cows are still milked to exhaustion, with little for their calves, even at this so-called cow shelter.

Besides dairy, the gaushala sells cow dung and generates biogas from cow manure. They also sell "Nectar of Urine." Mr. Mahavir says it's supposed to have curative properties for a wide range of ailments from diabetes to cancer. (One company is making a soft drink out of it—*Cowca Cola*.) He takes us to a small shed that looks like a makeshift chemistry lab. The energy from the biogas plant boils urine in a large glass beaker. Vapors are diverted into a pipe where the condensation takes place, and the distilled liquid flows for collection into a stainless steel vessel. Mr. Mahavir takes a cup of the final product and holds it up in the air. He tilts his head back, opens his mouth, pours, and swallows. "Ah," he exclaims.

We walk by another canopied and fenced-in area that traps and chains the cows for several hours a day to collect their urine. Inconvenience is not regretted.

Early in my engineering career, I created structural slope stability models of landfills. I was charged with assessing how they would perform during an earthquake. Garbage was an input parameter in my analysis. As a colored layer in a two-dimensional trapezoid on my computer screen, it seemed benign. The landfills expanded, and our workload increased. I overheard a colleague talk about this influx of landfill work as revenue. "We'll keep milking her for all she's got."

What strikes me this time when I return to India is plastic. Discarded plastic is everywhere, even in discarded cows. The mortality rate in most gaushalas is high. At Gopal Gausadan, three or four cows die each day, Mr. Mahavir tells us. When autopsies are performed, almost all the cows are found to have plastic bags lodged in their stomach, from foraging on the streets. The plastic gets stuck in their rumen, and after a while they can no longer digest other food, and eventually waste away and starve to death. The cows have become landfills, and they are still being milked for all they've got. They enter the gaushala, a place for refuge, as their bodies have become a place for refuse.

"Do you want to see a very sick cow?" Mr. Mahavir asks in his simple, direct English. "Almost at the verge of death."

We are taken to the quarantine area. Lying on the floor are two cows next to each other—one black, one white. Next to them are a mother and child. The mother lays on her side, while her child fits under her belly. Flies hover and land on them. The pink of their flesh is exposed. Polyethylene strips that once littered the landscape now clog their rumen. Skin and bones only, they can easily be mistaken for carcasses. Only their eyes still move.

"Do you ever euthanize them?" I ask, wondering if it would lessen their suffering.

"No," he says, bewildered. "Why would we kill a cow?"

They will die an unnatural death, naturally.

# THE FIRST CONCEPT

Overheard in Bryant Park in New York City: "In English cows go *moo*; in Italian, cows go *maw*."

I am reminded of Theresa Hak Kyung Cha's *Dictee*. Cha writes: "Mother, my first sound. The first utter. The first concept." Do the young Indian calves call for *Amma*? I had read *Dictee* in a class with the poet and memoirist Meena Alexander. At the time, I was making connections between what I've been learning about languages, loss, and animals. In most languages, the name for *mother* has the letter *m* in it. Human babies are only capable of vowel and nasal sounds (*m* and *n*) because their larynxes have not yet descended to pronounce the other consonants. The raised larynx allows babies to breathe and swallow at the same time, like chimps do. *Mmm* is the sound that can be said while nursing. It is the first utter. Mother and language are linked. "Mother tongue is your refuge. It is being home. Being who you are," Cha writes.

I think about my own submerged mother tongue, Tamil. The first concepts that attracted me to the Tamil poets were their articulation of an ecological grammar, an awareness of the grief of animals, and the understanding of the connection of inner and outer worlds.

I was reading Gay Bradshaw's book *Elephants on the Edge*, which explores post-traumatic stress disorder in elephants, when I took Meena's class discussing *Dictee*. "The rupture of elephant lands, lives, and history is mirrored in the legacy of rupture in their brains, bodies and behavior," Bradshaw writes. At the time I was also reflecting on how so much of my writing and activism around animals is related to the rupture between mothers and their young, like my small smalls in Cameroon, and the dairy cows in India. Who is responsible for the suffering of these mothers? "A matriarch's death means not only the loss of a loved one but also a loss of cultural and environmental knowledge," Bradshaw notes. How much do we all suffer from the suffering of mothers?

With Meena, I talk about narratives of trauma and how they have ruptures on the page. I wonder how to confront trauma without invoking it. Can the writer provide a type of sanctuary while exploring difficult material so that the reader can be transformed by it?

# RUMINATE

How do I tell you a story about cows in India? Cows who are worshipped. Cows who are weaponized, where violence is invoked in their name. Cows whose bodies are controlled by a caste system and commodified by an economic one.

Cows who are viewed as mothers. Cows who are forced into motherhood. Cows who are prevented from mothering. We take too much from mothers. We deny them too much.

I think about the Jain concept of saptabhanginaya—*sevenfold of truth*:

1. In some ways, it is.
2. In some ways, it is not.
3. In some ways, it is, and it is not.
4. In some ways it is, and it is indescribable.
5. In some ways it is not, and it is indescribable.
6. In some ways it is, it is not, and it is indescribable.
7. In some ways, it is indescribable.

This is what writing about cows in India feels like.

Some friends in the United States will be surprised when I tell them what I am researching.

*Isn't India a vegetarian country?*

In some ways it is. In some ways it is not. It boasts the majority of vegetarians, but it is not a vegetarian majority. Sometimes it is like explaining what Gandhi wrote in his first articles in *The Vegetarian*. Not all Indians are Hindus and not all Hindus are vegetarian. But not all meat-eaters eat meat every day.

*But there is no cow slaughter, right?*

In some places there is. In some places there is not. Slaughter is legal in a few states and illegal in others. What is legal or illegal, in some ways, is not a good indicator of what does or doesn't exist.

Some cows are killed because they are not considered cows because they are bulls or buffalo.

India can be both a country with lots of prohibitions on slaughter and a leading producer of beef.

*But milking doesn't harm the cows?*

In some ways it does.

In some ways it is indescribable.

Or it is describable but not described.

In the lineages of ahimsa I have been searching, concerns around dairy are mostly absent. Even the wisdom of the *Tirukkural* does not contain a couplet like this:

*Know milk for an animal's sore that it is,*
*And you will not drink it.*

Consider this account from an animal activist named Amit Chaudhery, documented in Yamini Narayanan's book *Mother Cow, Mother India*, about how cows are tricked into letting their milk down.

> At birth all calves are separated from the mother and allowed just enough milk to survive if they are female and

none if they be male. Enterprising cowherds sever the heads of calves and tie them just near the rump of the cow to create the illusion of her calf as she turns to watch while they milk the udders.

*What about ahimsa?*
In some ways, vegetarianism can be rooted in nonviolence.
In some ways, vegetarians can be complicit in violence.
In some ways, vegetarianism can be embedded in the violent hierarchy of the caste system.

While ancient Hindu scriptures introduced ahimsa, it was Jainism and Buddhism that popularized this concept and challenged both animal sacrifice and the caste system within Hinduism. Hinduism was later inspired by the vegetarianism of these younger religions.

As B. R. Ambedkar noted, "It is true that Hinduism can absorb many things. The beef-eating Hinduism (or strictly speaking Brahmanism which is the proper name of Hinduism in its earlier stage) absorbed the non-violence theory of Buddhism and became a religion of vegetarianism." Yet, he laments, "there is one thing which Hinduism has never been able to do—namely to adjust itself to absorb the Untouchables or to remove the bar of untouchability."

*So, people kill people over cows?*
In some ways, so-called cow protection has been used to beat and lynch humans.

In some ways so-called cow protection has mobilized violence against Muslims and Dalits suspected of carrying beef.

I say *so-called cow protection*, because in some ways it is not about protecting cows.

Cows are centered and decentered at the same time.

To respond to so-called cow protection that is not cow protection but a kind of terrorism or fascism, beef festivals pop up.

Eating beef can be a rejection to the violence of the state.

Not eating beef or drinking milk can also be a rejection to the violence of the state.

Sometimes refusal to participate in a violent system that harms both cows and people is the response.

One year, four Dalit men in Gujarat were flogged for skinning a cow. Dalits in that state mobilized a boycott, refusing to dispose of cow carcasses.

I return to the Bhakti poet Ravidas's quest for the city without sorrow.

Is this what it means to be a tanner set free?

Is Begumpura a refuge and a refusal?

*Did you ever find those sacred cows?*

The treatment of sacred cows in India is itself a sacred cow—too highly regarded to be questioned.

In some ways it is indescribable.

I remember this line from a character in Karen Joy Fowler's *We Are All Completely Beside Ourselves*:

> The world runs on the fuel of this endless, fathomless misery. People know it, but they don't mind what they don't see. Make them look and they mind, but you're the one they hate, because you're the one that made them look.

*I made them look.*
*I make you look.*

I return to the couplets of *Tirukkural*:

> What use is a raga that cannot be sung? Or eyes without sympathy?
> The jewel of the eye is sympathy. Without it, Eyes are but sores.

"Eyes are sympathetic," as my father would say.

# LANDSCAPES OF GRIEF

IN MY LIBRARY BUILT ON TOP OF A WATERLESS RESERVOIR, I learn about thinai, the landscapes of Tamil Sangam literature: Kurinci (*mountainous land*), Martha (*agricultural land*), Neythal (*coastal land*), and Palai (*dry land*). In her essay, "Retrieving the Margins: Use of Thinai by Three Contemporary Tamil Women Writers," scholar Chitra Sankaran tells us that thinai "[combines] geography [. . .] with the rules for poetry," and "is a concept that holds the ecosystem in an integral embrace."

Each physical landscape invokes an emotional one. The forest is for waiting, the coasts are for pining. Lovers unite in the mountains, fight in the farms, and separate in the desert. Sankaran says thinai is "a gestalt, a field of co-constitution in which woman, man, land, deity, flora, and fauna occupy a nonhierarchical relationship and form a continuum."

What is the thinai of the dairy, the piggery, the hen house, the slaughterhouse?

What does it mean if we no longer understand the grammar of landscape, or if our landscapes no longer contain the grammar they once held? Are these new combinations and intensities of fire, wind, and water, an evolving grammar?

What landscape, if not our bodies, can hold all this grief?

# MY KALYANAMITRA

2013. It is the beginning of the monsoon season in Rangoon. The air is burdened with moisture. NiNi has her notebook out, and I have mine. She and I are both civil engineering women whose works are governed by water. NiNi reminds me of my Uma Aunty, with her warm, round face.

She describes herself as a water professional by training and conviction. "Water World is my birthplace," she says. Myanmar, she tells me, is her "motherland."

Like a lighthouse, her motherland beckoned her to return in 2011, after many years working internationally. She wanted to bring water governance to Burma. It used to be if you said the words *research* or *environment*, you'd get thrown in jail, she tells me. Before, the lighthouse was warning to stay away. Danger. But there was a glimmer of a new light, that things could change. She arrived at a moment of shift. She wanted to make sure water and environmental issues were not excluded from the social and economic reforms that have been initiated. I read about her work on water resources and reached out hoping we could connect once Wan and I arrived in the summer.

I begin to tell NiNi about New York City, and how we receive water from three separate watersheds—the Delaware, Catskills, and Croton. I start sketching our reservoirs and aqueducts, mapping how they reach our city.

NiNi, then, starts sketching too.

"If I can sketch Myanmar, it looks like a kite," she says, as she draws in her notebook. "Like this. Like a cross. This is north. This is south. This is Thailand," she says, pointing to the east, and "this is Bangladesh," she says, pointing to the west. "This is the Bay of Bengal; this is the Andaman Sea."

"You know Kipling?" she asks. "Kipling in the old days wrote about the 'Road to Mandalay.' That is not a road, but a waterway. The Road to Mandalay is the Irrawaddy."

She draws a line down the middle of the kite.

From the archives in London, I was able to map Thatha's posts in Burma, from Myitkyina down to Tavoy, from Sittwe to Pegu. So many of Thatha's postings fall along the centerlines of her kite. Working on roads, bridges, and lighthouses, he too was always in proximity to water.

NiNi and I talk about the duality of names. Burma and Myanmar, Rangoon and Yangon, Tavoy and Dawei. The Irrawaddy is now called the *Ayewaddy*. She tells me I need to search both names, to get double the information.

———

The weight of the afternoon rain causes the tarp above the outdoor section of the Feel Myanmar restaurant to sag. A waiter pokes the plastic roof with the legs of an upside-down chair to direct the water to cascade down the sides. The chair leg–induced waterfall reminds Wan of the Korean expression 장대 비. *It's raining poles.*

At the café, the free Wi-Fi signal is slow. "When the rain comes, the signal is broken," NiNi says.

"All the schools in New York City have one company that serves them. Every time it rains, the internet gets slow," Wan chimes in, referring to the high school in the Bronx where he teaches.

Our first purchase in the country was a wide, green-patterned umbrella from Bogoke Aung San Market. I remember reading about Gandhi's trip to Burma in 1929, when my Thatha first met him. In his speech at Paungde, Gandhi asks the audience to "discard all foreign fineries," and remarks "what contrast foreign umbrellas presented against the picturesque Burmese ones."

I brought my green umbrella to the gates of the National Archives, an old white colonial building. At the entrance, Wan and I are told to wait for someone to come. A young woman arrives and asks if I have a letter of invitation. I show her a letter of award for a literature grant I received to facilitate this research. She scrunches her face in dismay. She does not deem my grant letter sufficient for entry. She calls the director to meet us. He looks me up and down and asks why I came.

What was I looking for? When I used to supervise geotechnical engineering drilling crews in the San Francisco Bay Area for water infrastructure projects, random passersby would stop and ask, "Drilling for oil?" The driller would laugh and say, "Drilling for soil!" But here in the archives, what am I drilling for? What am I trying to unearth? I can't easily say. Maybe it is because I don't know what I may find, or how it could be important to me. But maybe I do know, but it seems silly to admit, or perhaps impossible to find in an archive.

A monsoon is caused by a shift in wind. What was the shift in wind that turned Thatha into an activist?

NiNi understands this shift. She, like my Thatha, returned to her birth country during a period of growing change. In her case, it was after decades of fear and repression, which haven't fully subsided, and will return in full force in the coming years.

I am searching for something akin to her water conviction, to Thatha's water divining, and my own calling. This search is never direct, always meandering, and I often follow anything loosely adjacent.

I don't say any of this to the director. Wan and I stand under my green umbrella. The director looks at Wan and asks what he is doing here. I say he is my husband. He asks if he is also Indian. "No, Korean." Then both the assistant and director general become very curious and ask if we have children. I shake my head. He says to send him a picture when we do, because they want to see, and he writes his name down in my journal. In addition to thirty dollars and a handwritten note of interest, this is my price of entry.

We walk upstairs, setting our shoes and wet umbrella outside, and walk barefoot into the archives. It is a small reading room, with a couple of tables and a few desktop computers, which are fully occupied when we arrive. A librarian brings me a manual index to look for reports until a computer becomes available. I am hoping to get a hold of files of my grandfather's that were mentioned in reports in London.

Back in the Asian and African Studies Reading Room at the British Library, months earlier, I found myself riveted by a revenue report: "The most noteworthy event in the history of the year was the total destruction in 1877 of the Krishna Shoal lighthouse, on the south coast of Pegu." I found this sentence, which reads like the opening line of a novel, in the 1877–1878 report from the Superintendent of Lighthouses in British Burma.

Another noteworthy event that year, though apparently not the most noteworthy, was the murder of Mr. E. R. Woodcock, "the only European lightkeeper stationed at Table Island lighthouse," by one of the "menial employees" there. There were civil engineering and criminal mysteries to unravel in this tale!

I was curious to know if similar accounts existed for public works projects in Burma in the 1920s and 1930s. John O'Brien, the archivist of the India Office Records at the British Library, pointed me to an index of the Proceedings of the Government of Burma, Public Works Department, which contains reports of correspondences until 1924. I requested the volumes from 1919 to 1924. In each of these volumes, there was a section called *Establishment*, which included notes about engineer appointments and transfers. In the Establishment section, I discovered my Thatha. First, when he is appointed to the department: "Narayanan, Mr. M. S. Appointment of as a Temporary Engineer." Next when he is transferred to a different post: "Narayanan, Mr. M. S. Assistant Engineer. Order transferring from the Pegu to the Maritime Circle." Later, when he is granted "an extension of time in which to pass the lower standard examination in Burmese." And then again, regarding a nominal raise: "Narayanan, Mr. M. S Assistant Engineer Superintending Engineer, Maritime Circle letter No. 4010-14E (N. 10) dated the 6th May 1921 recommends that sanction be accorded to grant of the first annual increment of Rs. 20 per mensem to the pay—with effect from the 1st February 21."

The British Library did not have the records contained in the file number referenced in these reports. These proceedings were often copied to London, but the enclosures referenced were not sent. It's possible the records were still in the Rangoon archives, if they survived World War II, the British librarians told me. Would there be such riveting accounts from Thatha's posts in Rangoon?

Here in the Rangoon archives, I do not have better luck finding Thatha's records, despite my promise of (a photo of) my firstborn child. I do find some reports on the regulation of rivers without embankments, a register of bridges, and various reports on the bureaucratization of the capture of wild elephants.

I know that in 1928, Thatha was in Rangoon working as an engineer at the "New Lunatic Asylum" in Tagadale. What I can find about Rangoon's first "lunatic asylum" from the Yangon archival records is this: "Rangoon lunatic asylum situated at the northwestern corner of central jail, on open space, with good natural slope, affording excellent means of conveying the surface water from grounds during the rains." Of the people, it reads: "As might be expected, coolies, cultivators, and petty traders supply the greatest number of patients."

I wonder—How much of the madness was caused by the state? These bureaucratic reports speak more beautifully about the drainage of the asylum site than they do of the inhabitants.

In 1930, two years after my Thatha's posting at the asylum, Rangoon is filled with bloodshed. During the conflict between Indian and Burman dockworkers that is called the Rangoon Riots, the Indians barricade themselves at home and some even lock themselves into the Rangoon "Lunatic" Asylum. I wonder if and how the news reaches Thatha, who has taken thirteen months of leave during that time. Does this event, a year after Gandhi's visit, further shape Thatha's decision to leave? When he does move back to Tamil Nadu for good and starts his own experiment in Kallakurichi, it was not an attempt to barricade himself from the ills of society. Kallakurichi had no locks on doors or glass in windows. It was an experiment in trust.

———

When the Rangoon archives close that day, Wan and I stand under the awning of the neighboring Indonesian embassy until the rain stops, and walk three miles, through People's Park and down the steps of the Golden Pagoda. We navigate the uneven and flooded sidewalks of Yangon, and hear the beautiful call for prayers from the mosque. We arrive at our guesthouse, where all the young women watched K-dramas and want to practice their Korean with Wan.

That next morning, NiNi picks us up from the hotel, and we accompany her to various water and sanitation meetings and go for lunch at a café. She continues to sketch the Irrawaddy River basin in her notebook, and she tells me that land erosion, illegal logging, gold mining, and electrofishing are the major threats to this area, home to twenty-four million people. There are only fifty-six river dolphins left, all found in a small, protected sanctuary along the river.

She orders a bunch of vegetable side dishes for us to share—collards, okra, and eggplant.

I learn the Burmese phrase *thek that lut*, to request vegan food. I assume the translation meant something related to vegetables, so I always confirm that food has no fish sauce or other animal ingredients. The response is always confusion, something like, "Of course not! It's thek that lut!"

Years later, I email NiNi and ask about the translation, when I am in Korea visiting Wan's parents at their farm in Dunnae. She happens to be in South Korea, too, at a water conference at the time, and responds:

> The Burmese phrase for vegetarian is "thek that lut" 1. the first word "thek" ends with "k" sound means "life" 2. the second word "that" ends with "t" sound means "kill" or "killing" 3. the third word "lut" means "free from or exception." So, the overall suitable translation may be "food without life-based materials" (or) "meal without killing." or any good combination of words which you can deduce from 1, 2, 3 meanings.

*Thek that lut* conveys the intentionality and philosophy of veganism that extended beyond food. It wasn't a diet or fad; it was an ethical way of life. Thek that lut, *free from killing life*, echoes ahimsa, *the absence of a desire to harm*, and Jeevakarunyam, *compassion for all beings*. These are the ideas that are part of the love story I have been searching for.

NiNi asks how we survive in NYC as vegans. "It's surprisingly easy," Wan says. He also tells her about his parents' farm in Korea, where they grow black raspberries and have a large vegetable garden.

"You know what I thought? I was thinking you were Mongolian," she tells him. We talk about Mongolian ancestry throughout Asia, and how babies are born with the Mongolian mark—a blue butt.

"Koreans get it, too," Wan says.

"In Myanmar, some," she says. "Sometimes they get it in the wrong place, in cheek, forehead, and it disappears after one year."

I think about these disappearing marks of ancestry. What are the things we are born with but lose? When the mark disappears, does the body still remember?

The food arrives, and we start talking about the rehabilitation of the Four Rivers Project in Korea. "I follow everything," NiNi says. Wan tells her about his dad, who recently biked along all of the four rivers. He has a stamped river passport.

How many stamps would be on NiNi's passport of rivers? What about mine?

We return to discussing water and sanitation. She tells us about the sewage disposal plant behind 50th Street that goes into the Yangon River. Wan and I stumble upon it later that trip when we look for the Rangoon Lightship. "Hey look, it's their wastewater treatment plant!" I exclaim.

"Of course, it is," he says. Inevitably, we always find the local wastewater treatment plant on our travels.

My notebook fills up with numbers and units. Millions of gallons of water or sewage, miles of river, megawatts of energy.

"All my energy and time, I want to do for water," NiNi tells us. "I think it will really do good for our people."

She talks to us about all the challenges from both big logging companies, but also poor fishermen in deep poverty. Sometimes they put dynamite in the water to stun the fish or use cyanide, and it poisons them. "They didn't mean to harm anyone, including themselves," she says.

A big economic forum is happening when we meet. World leaders from many countries are looking for markets and resources.

"Madeleine Albright is here drinking Coca-Cola," NiNi says. Coca-Cola is looking for the Gold Rush, she thinks. "Chelsea Clinton was here and donated water bottles. Are we to clean them up, too?" she wonders.

Later, in our hotel, I read *The Economist*'s "Special Report" on "A Burmese Spring," about a country on the brink of change. I am reminded of when I visited Rwanda in 2005. Then, too, leaders in the United States and UK were swarming, and arrived by coincidence at the same time as us. We learned that Laura Bush and Cherie Blair met the same gorilla group we did. Later that month Bill Clinton visited Rwanda too. It was under his watch as president that the genocide happened. In 1998 he came to Rwanda, but never left the airport. His "apology," if you could call it one, was: "We did not fully appreciate the depth and the speed with which you were being engulfed by this unimaginable terror."

One day, Rachel and I met these Congolese tobacco businessmen, who were staying at our guest lodge. On a napkin they drew a map of the Democratic Republic of the Congo, and marked up Israeli diamond mines, South African gold mines, and South Korean logging concessions. It reminds me now of all the global forces trying to map and stake out their claim to Burma.

When I took dowsing lessons, my mentor told me about geomancy, mapping the energy of a place, including sites of atrocity. There are so many ways to map a country, every country.

What happens when we map water?

NiNi hopes to provide a voice to those in the Ayewaddy River basin. She holds consultative meetings and brings people from riparian towns and villages. She pays for their travel and accommodation. They have time slots to talk about their experiences. She gives them cameras to document photographs. They set up a village library and donate books. She tests their water quality. "Fourteen parameters. Turbidity, chlorine, BOD, VOC, temperature, everything," she says.

"We only need two things—understanding and participation," NiNi says, "for water governance to thrive."

I love this phrase—*water governance*—and it resonates with explaining our water infrastructure projects to the public for their input. In New York, we have one aqueduct that is leaking up to thirty-five million gallons of water a day. To fix that leak, they are designing a bypass tunnel around the leaking portion, but to connect the tunnel to the aqueduct, they would have to shut that aqueduct down for several months. This was a huge planning effort. In the past, I say, there wasn't the political will to consider this shutdown. "Because it will shut down their careers," NiNi astutely jokes.

She says there is also leakage from the reservoir to Yangon City. "Guess how much?" she asks. "Sixty-five percent!"

We talk about losses in our reservoirs and pipes. How do we estimate this unaccounted-for water? How can we truly calculate our losses?

≈≈≈

At the northern tip of the kite is the source of the Irrawaddy, where the Mali and the N'Mai converge at the confluence in Myitson. A Chinese power company aims to put a hydropower dam there, NiNi tells us, but it has been stopped so far. She looks for alternatives and talks about pumped storage hydropower. "It's completely outside the river system," she says. "We don't really dam the river, that is the beauty."

I ask where else they had done this. "It's useful in arid regions," she says. "Former USSR is using it. Pakistan. Anywhere that ends with a *-tan*."

Wan and I fly up north to Myitkyina later that month. We rent bikes from our guest lodge. The brakes don't work. They instruct us to stop with our feet. They mimic wheels turning and putting their foot down. "Okay, stop! Okay?"

We bike and see a large reclining Buddha, the local teacher's college, and a view of the Irrawaddy. It is Father's Day, and I am thinking of my late father, and wondering what he would make of me in Burma, tracing the footprints of his father.

We hire a car to get to Myitson. The power company has relocated many families near the proposed dam, and we drive by their new houses. We learn their new lands are not as fertile, and the animals return to the old lands to graze. But when the villagers go to fetch their wandering cows, they are not allowed back, because that area is now restricted.

The last stretch of the road to the confluence is a mixture of pavement and exposed stone. It smells like burning rubber and cow dung. The monsoon has not yet arrived in these parts, and the depth and width of the river shows its tardiness. In the confluence, we see two women and a baby boy bathing and laughing, a pack of water buffalo cooling off, and a few villagers panning for gold. I wade into the Irrawaddy and hold the river water in my cupped hands, raise it over my head like an offering to the sun, and then pour it over my body. I pick up the smooth, rounded stones from the riverbed.

When I meet NiNi again later, she tells me she keeps a bowl with stones from Myitson. "Do you know why the stones are round?" she asks. "They meet and fight and the sharp edges become round."

If the stones are fighting, I think water is the mediator, softening the edges.

---

At the western point of the kite, water is perhaps closer to an indicator species, a warning of more dangers to come. When we arrive in Sittwe, it is a disaster zone. Aid workers and water flood the streets.

This is where Patti first arrived in Burma after taking a steamship to Chittagong. Thatha was stationed in the Akyab Division in the early 1920s. He had just come from the Rangoon Lighthouse Division. Wan and I rent bicycles again and bike along the Kaladan River to a viewpoint where the river meets the Bay of Bengal. We see a lighthouse on an island in the distance, and a newer one on the shore with a truss frame. A skinny, shirtless man wearing a purple-and-blue longyi comes out to talk to us. He is the lighthouse keeper. He doesn't speak much English and mostly gestures. That lighthouse island was built by the British but was bombed in WWII. The newer lighthouse was reconstructed onshore.

We walk along an outcrop of siltstone or shale stone, and see a dead cow. There is a feeling of precarity in the air that I haven't felt elsewhere in the country. Maybe it is the heckling and harassment from young men, as we pedal back from the water, or the presence of armed guards around the mosque. When we come back to Rangoon, we see the *Time* magazine cover story of "The Face of Buddhist Terror."

In Sittwe, fruit bats sleep suspended upside down, hanging like fruits themselves from the trees. At dusk, when the last glimmers of light fade, they wake, spread their wings, and fill the sky. What have these bats witnessed over the past few years?

I arrange for a boat to take us to the lighthouse island. In the morning, we take a taxi to the boat jetty. The boat is late and getting fuel first. We wait at the jetty, but the tides keep coming in. Pigs are foraging, dogs are roaming. An outdoor café next to the jetty starts to shift all the chairs further inland. We give up waiting and go back to the hotel. There, we receive a call that the boat is broken, but now fixed, and that we should return to the jetty. By the time we get there, the water is up to our knees. We wade in this water to climb onto a big ship, to then jump into a smaller motorboat in the river. The boat is covered with a thatched roof, and side windows are draped in a tarp.

Two red plastic chairs are set up for us in the middle of the boat.

We leave at 10:09 a.m. The ride is choppy. The rains come down hard and we move against the tide. The young man on the boat gestures a rocking boat with his cupped hands and says, "No good." We turn back by 10:27 a.m. Our motorboat pulls up next to a ship filled with passengers.

The boat sloshes back and forth, and it is hard to stand upright without holding on to something. Again, we are instructed to climb onto this other ship. This time, everything is rockier and more unsteady. If we miss or trip during this boat transfer, we could fall and get crushed between the vessels. Slowly, we spot each other and make our way from the boat to the ship, swaying back and forth, and making our landing. When we get onto the large ship, it smells like urine.

They are preparing to evacuate everyone. Two wooden planks are placed to serve as ramps off the boat. One plank has step grooves on it, and the other is flat. Wan takes my hand and walks in front of me down the grooved ramp. An older man takes my other hand and walks down the flat ramp next to me. We three get down safely, but a few seconds later, one of the planks collapses. A lady, the baby wrapped around her, and the bags she is carrying all fall into the water. We hear her shriek, and then realize it is a roar of laughter, and everyone joins in.

We never make it to the Sittwe lighthouse.

In the center of the kite in Toungoo, the sun's rays burn our skin as we bike along road bridges on the Road to Mandalay. I buy some sandalwood paste—thanaka—and put yellow circles on my cheeks like the Burmese women do, to protect them from the heat.

A few days later we are in Bagan, the land of a thousand pagodas. Here, too, the rains are delayed and there is no relief from heat. The skin on Wan's arms starts to bubble, something we haven't seen happen before. It is like a thousand pagodas rising, and then disappearing like a Mongolian mark.

When we arrive at the eastern tip of the kite to Inle, we are relieved to be with water again. When I think of Inle, I see yellow and blue:

the golden meals of shan tofu, curried potatoes, and mango; and our time on the lake. "Streets are made of water," someone says. Instead of stray dogs, there are stray geese.

On Inle Lake, the men stand on their boats and paddle using their feet as oars, stopping at the floating markets, floating tomato gardens, floating houses. We visit a cheroot factory, and Wan buys a small case of mint cheroots, a souvenir of our unvoiced private hope.

～～

To reach the southern tail of the kite, we fly in a tiny plane, as we do to the other kite corners, since the roads are impassable. We don't have boarding passes but are given color-coded stickers. The flight crew makes sure we have the right color for the given destination. Mine keeps falling off because of the humidity.

We start our travels here, too, with a bike ride. I think we are biking to the beach in Tavoy. We cross a bridge and see Dawei University and take a right instead of a left at a fork. Instead of the beach, we find ourselves in a tropical forest populated with giant Buddhas and a procession of Buddhist monk statues along a hillside. The monsoon rains wash over us, and it feels so refreshing in the heat.

We see a funeral pyre up ahead.

I pedal to pass them as fast as possible to give them their privacy. I think back to when we immersed my father's ashes in the river Ganga, and my disdain for the tourists watching us grieve.

Thatha's last post in Burma was in the Tavoy Lighthouse Division, where the twins Amirthalingam Uncle and Saraswati Aunty are born in 1932. I once marveled to Amirthalingam Uncle about how Patti safely birthed thirteen children. For the first handful of children, Patti returned to her mother's house in India for her delivery. But he and Sachi Aunty were born in Tavoy.

"So much was trusted to prayer and boiled water," he told me.

When Patti died, my parents came to the funeral carrying my older brother who was only a few months old. My cousin Lalitha told me how much my father fussed over my brother and worried for his health. "Hot water. Hot water. Hot water," my father kept saying.

Boiled water chanted is its own kind of prayer.

I, too, put trust in fire, water, and gratitude, and learn the words for *hot water* and *thank you* in the lands I visit.

*L'eau Chaude* and *Merci* in French.
*Maji Moto* and *Urakose* in Kinyarwanda.
*Sudu Thanni* and *Nandri* in Tamil.
*Ye·nwè* and *Jày·zù* in Burmese.
뜨거운 물 and 감사합니다 in Korean.

That first day, we were unable to spot the lighthouse in Tavoy, but when I went back to our guest lodge, I found the website of a lighthouse aficionado based in the United States. He posted a satellite image of a lighthouse which was located further down the Tavoy River, at the mouth.

I take a screenshot of the aerial map and ask the lodge if we could get a boat to take us from Dawei, along the coastline around the Andaman Sea. "Danger" is the one-word reply. I don't know what they mean and whether *danger* refers to water or to pirates.

Instead, we hire a driver to take us south down the mountainous coast, until we get close to the lighthouse island.

Our driver, U. P. Aung, glides us down this beautiful landscape. We stop at a pagoda waterfall. He buys us sodas. We drive on mountainous, red, rusty roads. Children bicycle in their green school uniforms, while gracefully balancing umbrellas with one arm to shield them from the sharpness of the sun's rays.

We arrive at this small village on the sea where we can see two islands in the distance. We don't yet know how to get there. We stop by a small hut. Our driver asks a woman in the village for a thimbaw.

The word for *boat* is the same as the word for *papaya*, that green vessel carrying orange flesh. She grabs an umbrella and takes cover from the sun as she walks toward the water. She comes back and tells us there is a school in this village, in Launglone Township, and the headmaster's brother has a boat, and he agrees to take us. Three men accompany us: the headmaster's brother; a younger man; and an older man, whose foot is covered in white gauze.

So much of my searching, dear reader, has been like this. Having a small scrap of intel and following it to the next place, not knowing who or what would await me there.

Wan and I roll up our hiking pants, take off our shoes, and wade into the water and climb into the boat. The most gorgeous blues of water blend into the blue-and-white skies. Minutes pass and I can't believe our luck, arriving here with only a screenshot of Google Maps to guide us here, and the generosity and kindness of strangers.

Minutes pass and I don't know who or what we will find on this island, and if they will let us in. We approach the southern coast of the island and then sail up the east side, looking for an entrance. We spot a sign midway up the east side of the island. We notice a small shack and approach the shoreline.

A dog is the first to greet us from the thimbaw. Wan and I jump off the boat in the shallow water and wade to the shore, carrying our shoes. The dog leads us to three other people. Two lightkeepers, one older and one younger, and the younger man's wife, Kay Kay. There is a sheet of corrugated metal that reads "Mibya Kyun" (*light island*) and "Reef Island," the other names for this rock.

"You want to see the lighthouse?" Kay Kay asks.

"Yes," I say. "Do you get many visitors?" I ask, wondering if others have come before me on a similar journey.

"No," she says. "You are the first."

She's a city girl from Yangon who speaks English. She has been married to the lighthouse keeper on this remote rock for the last two years. She asks how long we've been married. I tell her five years. She says it's time to make babies.

Kay Kay and the lightkeepers escort us up to the lighthouse, and as we hike our way up the island rock, mosquitoes devour our legs, and we look out for snakes. The younger lighthouse keeper finds one and relocates him off our path.

It is humid, muggy, and the trail is overgrown with green vines and shade. It takes about thirty minutes to make the trek. When we get to the top, the sky reemerges, and we meet this old square masonry lighthouse with a rounded top. It's not what I imagined. It is run-down and dingy, but the view is gorgeous.

It was built in 1883. We stand 299 feet above sea level. The light is white and flashes every five seconds. I have a sense of connection, knowing that Thatha and I both stood at this point, and share this same view.

I think about Thatha, contemplating his life and his future from the top of this rock. Is he watching the riches of Burma being shipped out from this post? Does the river or sea persuade him it is time to leave?

Next to the lighthouse sit these dilapidated concrete colonial housing quarters. Is this where Thatha and Patti lived? Could the river hear Patti and Thatha's eight children (at that point) singing their Tamil songs from this building? Kay Kay calls it the Monkey House, because the macaques have taken over. They eat coconuts and then throw the shells and have broken all the windows. Do they see their reflections in the broken glass? The monkeys have now displaced the human dwellers, a decolonization of sorts.

We thank Kay Kay and the lightkeepers for their hospitality. They ask us whether they can come back with us to the mainland. Their trips back to civilization are limited. They run into the shack and change clothes and board the boat with us. The dog is left waiting on the shore. I write in my journal that he looks sad as we sail away. When we return to the village, we drop them off in town and start our way back to Dawei. The rains, which have held out the whole day, now make their appearance. In the car ride, on the radio, I hear a Burmese version of Madonna's "La Isla Bonita." We found our own beautiful island, Mibya Kyun, and touched Thatha's work.

The car comes to a halt. We have a flat tire. Wan helps the driver fix the tire, but then the driver can't find his car keys. Darkness sets in along with the rain. Two ladies, two men, and a few others in the local village form a search party, carrying their umbrellas and flashlights looking in the road and drainage ditches for the lost keys. Wan finds them in the trunk. The driver hugs him, and everyone laughs. We encounter another vehicle stuck in the road. We offer to give the driver of that car a ride since his axle was broken and there is no way to fix it now. We arrive back at our guest lodge and feast on a dinner of tofu and watercress and roasted cashews. I write in my journal, "Best cashews ever."

~~~

I meet NiNi again in New York the following year, in 2014. She is attending a meeting on the role of women. She's leading an organization called Water Mother. In her speech, she notes that women's issues are often split into two categories only: "One, women are victims; and two, women are marginalized in terms of percentage [seats] in the parliament." She went on to say, "Please allow me to raise the flag of Myanmar Women in third dimension, which is 'women are self-reliant, determined, and leaders in sustainable water governance and development.'" She recognizes women (regardless of age, marital, or reproductive status) as water mothers—the environmental stewards they are and have always been—and empowers them as leaders.

We are in another restaurant, both of us again with our notebooks open and writing. I tell her about my work protecting endangered species in our watersheds, and my desire to create a strategic plan for animal protection in New York City, applying my work in environmental planning toward animals. She had just taken a post protecting wildlife in the Tenasserim region, where the long-tailed macaques on the lighthouse island live.

At this meeting in New York, she introduces me to a beautiful phrase called kalyanamitra. It means *spiritual friend or coworker*. How, when two people meditate together and sit side by side, it is more productive than each of them doing it by themselves. We feel strength and support, knowing someone else is there with us on our journey.

My dear reader, you, too, are my kalyanamitra.

# A LIVING LINK

WHILE KING ASHOKA ERECTS EDICTS ETCHED IN STONE throughout his vast empire, his daughter leaves behind a tree. Sanghamitra carries the southern branch of the Sacred Bo fig tree, where Buddha found enlightenment, from Gaya to Anuradhapura, in what is now Sri Lanka. That sapling is perhaps the oldest-documented planted tree, the only living link to the Buddha, and also to my namesake. It is Sanghamitra's final wish to be cremated in front of this tree.

Pilgrims now visit the tree and ask for babies. They ask for healthy crops. They ask to protect their fields from elephant feet.

If we were to meet, I would ask the tree what it was like to be split from her ancestors and take root in this new environment, an experience she shared with my namesake. Tell me, Sacred Bo, what do you remember about her?

In exchange, I would offer stories about another Sangamithra, from another time.

# TRYING

In the years I try to conceive, I spend my days worrying over New York City's water supply. We plan an ambitious project to repair one of its leaking aqueducts, which requires taking that tunnel, which supplies about half of the city's water demand, offline for up to eight months. There are many things that need to get done in advance. We have to wait until we are ready.

When I am finally ready, my body isn't. I arrive one morning when I am thirty-nine years old to the twenty-seventh floor of a building overlooking the Brooklyn Bridge to meet my first reproductive endocrinologist. She is sitting at her desk with diagrams of the female reproductive system. She walks through what is typical, and what happens when we age. She speaks fast and technically, like I do when describing watersheds and aging infrastructure.

At work, we discuss time and budget for our infrastructure projects. What can we afford and when? What should we prioritize? What are the impacts of our actions? My job is to direct, to project manage. I assign tasks. I problem solve. Most New Yorkers don't think about the water coming out of their tap and all the planning that goes into it. It is invisible labor.

There is also the invisible labor of fertility. I project manage myself. I pee on a stick and wait for a smiley face. I go to acupuncture. I take Chinese herbs to nourish ren and chong and mobilize essence. Try. Fail. Try. Fail. I drink raspberry leaf tea, eat fertility-friendly foods—seeds, embryo-looking beans (which Pythagoras avoided), and black- and blueberries. Try. Fail. I receive Maya abdominal massages. I drink more water and start running. Try. Fail.

"It will happen," my unconcerned soon-to-be-former gynecologist reassures me in my mid-thirties. When it doesn't happen by the next year or the following one, she assumes it is because I do too much.

On top of my day job, I am writing and editing the book-length twentieth anniversary edition of *Satya*. My doctor asks if there is anything I can cut back. She says I need to prioritize getting pregnant, as if I can just take my arm and in one broad sweep clear my desk to make room for parenthood. As if sufficient wanting is a precondition for pregnancy. She isn't the only one who wants me to prioritize. In writing circles, others advise that to finish my book, I should say no to time-consuming activist projects. I have to choose. Book over activism, or baby over book. My ambitions, they tell me, are my curse.

What they don't understand is that this is not about ambition. Writing and activism alone are insufficient to describe my desires, how I think and move in this world. How do I speak of a love for this planet so deep, I want to devote my entire life to its care? That this love preexists any possibility of a child. That this love defines who I am. That this love has animated all my quests. My desire for Kallakurichi, for losing cruelty, for a catena of kindness.

That it can't be dismissed as a failure of priorities. But I worry, too, that my brooding over the world is impacting my ability to brood my own eggs.

"Just relax," my gynecologist tells me. "Watch TV. Go into vacation mode," she says. (Note: If you want to upset a busy woman, tell her to *just relax*.) I stress out trying to de-stress. Try. Fail.

"Humans are not very good at this," my supportive reproductive endocrinologist tells me a year later in an attempt to destigmatize infertility. She lays out the statistics, saying that even in our prime childbearing years in our twenties, there is only a 25 percent chance of getting pregnant any given month. As we age, that drops to 10 percent. My job, my book, my activism, don't seem to play a role in these statistics. Still, the numbers are not reassuring.

When people ask me if I have children, I don't say yes or no, but rather, "I have a dog." When I am twenty-eight years old, Moo Cow jumps into our hearts like she does into our bed—an unassuming bundle who spreads out and occupies the entire landscape. We twist and bend to accommodate her with joy and devotion.

When my friends from high school start having children around the same time, we share stories—theirs human, mine canine. I delight in the similarities—"Moo Cow does that too!" I blurt, to their dismay.

Before I am trying to get pregnant, I am trying not to get pregnant. I succeed in this kind of trying.

I pursue an MFA in Creative Writing in my early thirties, while I work full-time planning and protecting New York City's watersheds. I write on my commute from Brooklyn to Queens on the F or G trains. I am always writing in motion, in thirty- to sixty-minute slices of the day.

One summer after my MFA, I attend a weeklong writing workshop in Provincetown, carving out time for myself and my writing away from work and away from the subway. I walk in the morning, discuss historical narratives in the afternoon, and catch the last glimmers of daylight before retreating each night to read. A fellow writer posts on my page, "Ah, life without children."

To be a woman of a certain age without children is to be both pitied and resented at the same time. I return to Provincetown the following year, and one of my classmates asks how I manage to be involved in so many things on top of my day job. Our writing instructor interjects casually and comically, "It's because she doesn't have children."

As someone who is both waiting and wanting, these remarks sting—like getting bit by an insect you thought couldn't hurt you. Sure, I know these comments are more about the commenter than about me, but I don't appreciate my uterus being the subject of public discussion. It is true, I waited in part because I feared having children would limit my ability to pursue my other passions. I am not sure if I can squeeze one more thing into my life and not lose myself in the process. One of my new writing friends, Mary-Beth, comes to my defense, saying many people don't have children and they aren't doing what I am doing.

In a dinner we share back in New York City, this friend tells me that I will continue to be who I am and find the time to do what I need to do, with or without a baby. Sure, it is hard, but isn't it already? It isn't only because I didn't have children that I became who I am. While I have been fighting the false choices others suggest I make, I fall victim to the same binary thinking. She points out the possibility and not the limitations. There is neither pity nor resentment for my empty womb, but a faith in me that I have failed to see.

New York City's drinking water supply reservoirs have become a haven for bald eagles. Because these birds are listed as a threatened species by New York state and protected federally under the Bald and Golden Eagle Protection Act, any time construction or maintenance is needed on our water infrastructure, we do everything possible to protect these eagles and not disturb their nests. There are imaginary concentric circles around each tree, 330–660 feet in radius, where work is prohibited during nesting season. It takes a village to protect a nest of this vulnerable species, but why did we wait until it was almost too late to start trying?

When Wan and I get serious about trying, we take a mini-vacation and rent a house in the Catskills near Ashokan Reservoir. Ashokan is named from the Iroquois for *place of fish*, but it always makes me think of King Ashoka.

One day, we hike the Jimmy Dolan Notch trail. I remember this name because we continue to curse it constantly. It is one of those hikes where every step is like a chess move, and you need to think ten steps ahead. Moo Cow, who is twelve years old at the time, still has this puppy zest with no patience for planning. She chooses the most difficult terrain to go up. We make it to the top and her energy wanes and she takes a half-hour nap in the sun. I worry about the descent, but Mookie flies down seamlessly. It is only when we return back to our home in Queens that we realize she is in pain. She has these tremors. Her body experiences a tectonic shift. Her legs do not receive signals from her brain, and her back paws "knuckle." A neurologist diagnoses her with herniated discs. Her demeanor changes from being a Tigger to an Eeyore in Winnie-the-Pooh parlance. During this period, I read *The Art of Stillness* by Pico Iyer. Our lives, once jolted with her intense joie de vivre, slow down with the same intensity. All I want is for Moo Cow to feel better and heal. *This could be enough*, I think. *The three of us. Why ask for more?*

Given my age, I no longer have the luxury of postponing. When I don't conceive, irrational fears flood my brain. Is this punishment? Am I being selfish asking for more? Am I being tested? Deep down, I know better than to believe such things, but can't stop thinking them.

Superstitions are appealing, because rituals accompany them. Rituals give you a sense of power in situations that are powerless. The opposite of superstition is science. After all my natural and holistic experiments fail, I seek out science to assist with reproduction.

When my reproductive endocrinologist explains there are medications to manipulate my reproductive system, my first thought is about animals and all the ways humans have manipulated their bodies: the forced impregnation of cows to produce milk, and the forced separation from their young; depriving egg-laying hens of light to trick their bodies to molt and start another laying cycle; the breeding of laboratory animals; and the oiling or shaking of geese eggs to suffocate the embryos, while the unaware mother goose continues to sit on her nest, waiting for her baby to hatch.

Unlike these animals, I am there by choice, but I still feel poked and prodded like the animals I have written about before. Several times a month, I wait for a nurse with an owl tattoo on her forearm to draw my blood. While science is the opposite of superstition, it, too, is full of rituals. I don't know the outcomes of all this labor. Of the cocktails of drugs and supplements, the pelvic ultrasounds, the whispering on the phone at work to nurses, insurance companies, and pharmacies, and the schlepping to various appointments. But I submit to these rituals.

Wan and I submit to ritual injections, the bruised bellies and thighs. He lines up all the medications on the bathroom counter and in a back corner of the fridge. He sets up his iPad and we watch and imitate the video instructions. I clean my belly with an alcohol swab, while he prepares the solutions, tapping the air bubbles out of each needle. To make a baby this way requires a different kind of intimacy.

Despite the submission to these rituals, my body does not submit. It refuses to be taken over. My follicles do not respond to the medication. Like a mother goose sitting on an oiled egg, I foolishly wait.

My body is at war with itself. "Disease is your body's illogical way of protecting itself," one nurse explains to me. I have largely managed my colitis well throughout the years with my vegan diet, but in my thirty-ninth year, when all of the trying is happening, my body flares up. By chance, I discover a clinic at Mount Sinai Hospital in New York City focused on IBD pregnancy and planning. "We are the only clinic in the world looking at the link between IBD and fertility," my new favorite gastroenterologist tells me. She is sharp and so full of curiosity about me. It is as if my body proves some of her suspicions regarding the role of IBD in ovarian reserve and response. She has a theory that, if we control inflammation in my body, my chances of pregnancy would improve.

I am the test subject for this theory. "I don't want you to feel like a guinea pig," my GI doctor tells me. I don't. I am happy to be making science—to potentially help women like me, so their failures aren't blamed on their success, so they aren't told to *just relax*. Again, I subject myself to a new ritual, three-hour infusions of an anti-inflammatory drug every two months to allow my body to heal.

I go on a writing retreat in Hawaii, and, on the recommendation of a fellow writer, I sign up for a Watsu—*water shiatsu*—massage. Humans are land mammals, but it is only in water that our bodies can truly free themselves of the tensions we carry. The massage therapist ties floatation bands around my thighs, and I begin to drift in a shallow pool of warm water. I trust in the process and begin to let go. I feel safe and unburdened and wonder if this is what the womb is like. Could my womb provide this kind of safety? Could my body, at war with itself, give this kind of peace? *The world is safe. I have you. I will carry you. You are cared for.*

The truth was that the world was not safe, and I was worried about the war on the outside, too—the destruction of our planet by our species. Is my worry on the outside contributing to the war on the inside? Could I create a sanctuary both inside and out? Akam and puram.

There was a solar eclipse crossing the United States on my fortieth birthday. Wan and I decide to take Moo Cow and drive all night from New York to the path of totality in Tennessee. It is a wonderful thing to have a cosmic event overshadow your milestone birthday. The whole country celebrates, and it invokes so much awe. I am not focused on the limits of my fertility or mortality, but rather the wonders of the universe. I feel very alive.

We meet our friends Rocco and Xenia at their campsite. We have our eclipse glasses on.

At first, it looks like someone took a bite out of the sun like an apple, then it morphs into a banana. The temperature begins to drop as a darkness sets in. It is unlike the magic hours of dusk and dawn. A gorgeous blue hue accompanies the fading light, and the corona of the sun behind the moon puts on a fireworks display. When it shines like a diamond ring, all the humans in the campsite start to *ooh* and *aah* and I cherish their candor. "It's f—ing beautiful," I hear someone say. The crickets start to chime in, too, and the bats and stars make an early appearance.

I am grateful to have taken this long trip with Wan and Mookie, and to share this memory. A few short weeks after this, everything changes. Moo Cow develops a brain tumor. We pursue radiation to shrink the mass and relieve the pressure. We submit to a whole new schedule of trying. Of early morning drop-offs at the Animal Medical Center on 62nd Street. Of evenings ascending from the bowels of the earth in midtown Manhattan from the F train and walking east until we reach the edge of the island to pick her up.

We submit to the steroids to control the inflammation. We submit to managing the effects of those steroids. The prednisone makes her susceptible to bladder infections. When we suspect she has a urinary tract infection, I dive under her with a clean plastic container while she urinates to capture her pee. I take a pipette, just like in high school chemistry lab, and transfer it to a sanitary vial to deliver it to the veterinary's office across the city.

There is a period when her UTIs are so bad, we have her wear diapers, launder them daily, and wake up multiple times in the middle of the night to take her out. After her walks, we lay her down on one of her mats that we started to call her Koala changing station, just like the diaper tables in public restrooms. Sometimes, she is too tired to move, and we carry her in her dog bed from room to room like a magic carpet ride. Her tumor makes her confused, and she is vulnerable to falling and getting stuck under furniture. We block off a section of the apartment for her to safely stay unsupervised when we are not there. We call it her Manhattan studio. She can no longer jump into our bed for comfort, so we start sleeping on the floor with her. It is a slumber party every night. She loses so much muscle mass in her back legs that we use a harness in the front and back to walk her. We walk as a team, with the person in the back prepared with a wee-wee pad to capture any accidents that could occur in our building's basement and avoid judgment from our busybody co-op board. Every day, several times a day, the three of us continued to navigate this ever-evolving dance.

The three of us together. Trying.

Some people believe that it's not until you have kids that you understand the depth of love. When there is a tragedy in the news of something that happens to a child, politicians speak as parents. Parents imagine if it happened to their kids. There is an assumption that having kids awakens a special empathy. I like to believe my heart aches just as much to the news of such loss. I know it has.

I also know that my life has purpose and fulfillment with or without children. I understand the choice not to have children and the choice to have them and that not everyone gets to choose.

The word *essay* comes from the French *essayer—to try*. You don't know the outcomes of your labor unless you try. It's easy to confuse trying with failing, when you have been doing it so long. But I've come to realize trying is far closer to loving.

# SANCTUARY CITY

ACROSS THE OCEAN, WHERE EARTH, AIR, AND SEA COLLIDE, a pair of Bonelli's eagles builds a nest in Cassis, on the southern coast of France. In the Calanques National Park, they can soar to the top of Cap Canaille and plunge, like the steep limestone cliffs themselves, into the Mediterranean. The Calanques is the youngest national park in France, and the only one within an urban metropole.

I have always been drawn to liminal spaces. I, too, live in a city with a national park—the Jamaica Bay Wildlife Refuge. My work inhabits this nexus of land and water, of urban and wild.

I arrive in Cassis in January 2018 for an artist residency, and to explore the possibilities and tensions of coexistence with other species in these in-between spaces. When I received the call that I was accepted to this residency, several months earlier, I was on the way to the vet to pick up Mookie. It was the day they confirmed she had a brain tumor. She made it through her radiation course over the next two months. Wan encouraged me to go, while he stayed home with Mookie. We would FaceTime and have meals together during this month apart.

The day after my arrival in France, I take a walk just before sunset and see a crowd forming at the marina in Cassis.

"Qu'est-ce que c'est?" I ask a woman, in my limited French.

"Une baleine," she says. *A whale.*

I spot a gray-finned body, stuck between docked boats with an arc of red blood streaking the top of his head. I want to see wildlife on this visit, but not like this. A diver assists the whale, guiding him to sea, as a boat crew of local firefighters follow in an inflatable orange raft.

I ask one of the firefighters if the whale is injured.

"Oui mais n'est pas grave." *Yes, but it isn't serious.* They have tough skin, he explains. Another woman tells me the whale is not alone. "Il y a trois mais deux sont morte." The phrase echoes in my head. *There were three, but two are dead.*

How I am comprehending French at a level beyond my studies is a mystery to me. Maybe it's like an emergency, where a woman acquires adrenaline-charged superpowers to lift a car to save a child. Perhaps, in this whale crisis, my own French comprehension superpowers kick in.

The news later that night confirms two beached Cuvier's beaked whales. The male from the marina, and a female found dead outside the port. Cuvier's beaked whales normally travel in pods deep within the sea. Unlike Bonelli's eagles or myself, they would not choose to come near the shore. The piercing sound of military sonar—a phenomenon of increasing frequency—likely drives them here.

After the male is escorted out of the port, the inquisitive pod of human onlookers collectively sighs in relief. That night, however, I stare out my bedroom window at the blinking green glow from the Cassis lighthouse, thinking about this injured whale alone at sea, grieving his companion.

The next day, I ask an environmental police officer with the national park what happened to the female. He tells me they collected her body to do an autopsy.

Not wanting to dwell on this subject, he attempts a joke: "Whale became sushi," he says.

I do not laugh.

I recognize this instinct as a defense mechanism, a way of avoiding discomfort. I practice lingering in that liminal space between grief and comfort in search of a deeper wisdom that defies the initial response of avoidance or futility.

I turn toward the turquoise sea for its teachings.

~~~

If you dive into the Mediterranean, you will find other bodies affected by human activities. The neighboring city of Marseille historically was home to soap, soda ash, and lead manufacturing. These industries relied on the labor of immigrant bodies and resulted in the pollution of bodies of water and bodies in water. Mediterranean seagrass stabilizes the seabed, but still carries traces of metals produced over a century ago. The luminescent jellyfish can thrive in this environment. The increasingly threatened, water-purifying sea cucumber, what one researcher called "the medicine of the sea," is also present. Resilient bodies who survive despite adversity, they are also considered "ecosystem engineers," organisms that play a critical role in creating or modifying a habitat.

Perhaps I am drawn to these species because they remind me of New Yorkers surviving in our difficult environment. Like Marseille, my city is also recovering from a history of industrial pollution and is reimagining its relationship to nature. Perhaps I am also an ecosystem engineer. My work in watershed protection is intertwined with species protection.

~~~

Take. The Endangered Species Act defines *take* as "to harass, harm, pursue, hunt, shoot, wound, kill, trap, capture, or collect, or to attempt to engage in any such conduct."

Too often, humans modify the habitat in a destructive way. We *take* until it is almost too late, and protect species only on the brink of extinction. *Take* is the operating principle of both colonialism and capitalism. What if *take* and *no take* aren't our only options? What would it look like to *give*?

Consider the astragale de Marseille: On a "toxic tour" of the Calanques, I encounter this rare protected species, a short thorny plant growing in the cracks of the limestone cliffs. Because of this plant's shape and unpleasant feel, she has been given an unflattering nickname, *Mother-in-Law's Cushion* (coussin de belle-mère). The name fails to appreciate how remarkable it is that she can grow in this harsh landscape, where few other plants can. Her sharp features allow her to limit wind erosion and prevent the spread of metal contamination to the air and sea. As Isabelle Laffont-Schwob, an admiring researcher, explains to me, "We have to protect this plant, but she is protecting us." The astragale lays her body on the edge of a cliff to shield us from harm.

---

Scientists, researchers, and rangers are figuring out how to govern these lands and protect the species within the Calanques. The plants themselves have their own protection measures. Some have leaves that are hairy, spiny, or scaly. They hide in their own bodies. How can we feel safe in our surroundings?

My home city of New York is supposed to be a sanctuary city, welcoming immigrants, at a time when our president spews xenophobic vitriol and bans bodies at our airports. My borough of Queens is the most diverse county in the country, where eight hundred different languages are spoken. This area of southern France is also home to many immigrants from Africa and the Middle East. When I walk around Marseille, it feels like Queens to me, which is to say I feel at ease walking in my brown body, by which I mean I feel unnoticed, invisible. But is it possible to be both seen and at ease?

As a newcomer, I am told that Marseille is an ungovernable city. This seems to suggest different things to the different people I meet. To one young college film student, it means resistance. "You cannot have gentrification in Marseille," she explains. In other French cities, the poor have been relocated to the outskirts of the city, but in Marseille they are rooted in the center, she says. To an independent book publisher I meet, Marseille is both a colonizing and colonized city. It is a city that has to reconcile both histories. To the national park rangers I go on hikes with, it is a challenge to change the mindset of a populace accustomed to doing whatever it pleases to now respect the new rules of the park. This label of ungovernable can be taken as pride—a wildness that can't be controlled. But to others, this status creates a habitat for corruption and violence without accountability. Is *ungovernable* a racially coded phrase or an excuse for the limitations of government?

Outside my apartment in Cassis, Mexican agave and prickly pear cactus grow, a legacy of colonialism. They are viewed now as an invasive species, but let's not forget who invaded who.

One evening, the gardener Gilles Clément visits the resident artists. He dislikes the word *invasive*. He prefers the phrase *emergent ecologies*, and speaks about a commingling of species that has always existed with the migration of humans around the planet. I found *native* and *invasive* to be a troubling framework in environmental sciences and am grateful that Clément proposes different nomenclature.

I read that Clément "discovered" a butterfly in Cameroon. I ask him more about it that evening.

"Well, yes," he says. "It was a long time ago." This isn't a source of pride for him, but shame.

He was sent to Cameroon by a museum. He wanted to observe butterflies alive and see how they behaved, "by night and by day." But the museum wanted him to kill and capture them. When he gave the museum the two thousand dead butterflies, the response was, "That's all?" He no longer wanted to kill animals to study them.

It perhaps is his Ashoka moment, his radical turn, to study life and not cause death.

In this butterfly collection, there is at least one species that was unknown. "It was probably known to the Indigenous community," Clément clarifies. To his dismay, it was named *Bunoeopsis clementii*.

"The name does not give any description about the butterfly," he says. It reflects a fraught colonial relationship with other species.

I want to know more about emergent ecologies and what peaceful commingling could look like. This is the question I carry during my walks in the Calanques. Jeremy, a park ranger, tells me about the caterpillar processionnaire du pin. They are spiny and can cause a rash when touched. Park visitors often set them on fire. "It's not acceptable," he tells me. "We have to change the mentality."

On one hike, I see the footprints of the wild boar, le sanglier. A researcher with the park, Lidwine Le Mire Pecheux, describes how le sanglier should be examined in three different contexts: écologique, sociologique, and politique. "Pigs don't have problems with humans, humans have problems with pigs."

Lidwine, always speaking in threes, like a college professor articulating her points, shifts her attention to the fish mérou. "It is an emblematic species, a protected species, and an economic species," she says. Recently, the national park instituted "no-take zones" around the park, and the mérou were protected. But since their populations are increasing, she says the government is now considering removing the protections. They only receive a short respite from our harm.

Several archipelagos are part of the Calanques. Two of the islands have almost rhyming names: Riou and Frioul. Puffins from the Atlantic Ocean come to Riou to breed. While no humans are on this island, the pellets of seagulls are full of plastic, demonstrating our presence even in absence. Rats, cats, and humans inhabit Frioul, as well as some puffins. To protect the puffins, humans trap and kill the rats, but sterilize and relocate the cats to give them to an animal rescue group. Conservation is messy, and it is hard to tell the good from the harm.

In New York City, our parks department initiated WildlifeNYC, a campaign to recognize bats, beavers, coyotes, deer, hawks, raccoons, squirrels, and our other urban neighbors as fellow New Yorkers to promote coexistence. Notably absent from the campaign are the most iconic New Yorkers: pigeons, rats, and geese. Can the label *New Yorker* provide them sanctuary, too?

~~~

What is "The New Colossus" of our Invisible Animal City? What could be chiseled on our Statue of Animal Liberty?

> Give me your coyote, escaping the highways and malls of the suburbs.
> Give me your mother Canada goose who fiercely protects her young.
> Give me your ruddy ducks who mate for life in the drinking water reservoir.
> Give me your Harlem deer who chose to find shelter in public housing.
> Give me your pigeons, your rock doves, who dive-bathe into that parking lot puddle of stormwater with much delight.
> Who line up dutifully under the elevated LIRR tracks, as if it is where they were meant to roost.
> Give me your subway rat who knows better than the MTA when the train is coming.
> Give me your bodega cat.
> Your baby raccoons abandoned in a diaper box in Forest Park.
> Give me your bold bold squirrels, gray or black, taunting the dogs who will never catch them.
> Give me your dolphin in the Gowanus Canal,
> The humpback whales in the Rockaways,
> The oysters returning to Jamaica Bay,
> Give me all who came here by choice, chance, or circumstance,
> Who breathe life, save life, and protect life in this city that is so hard to survive.

I remember when they came for the geese in New York. It was a secret ops mission—hired contract killers who came in the middle of the night, zip-tied their legs together, and stuffed the geese into plastic crates. The job was timed with the summer molt, so the geese would be unable to fly away.

They were gassed to death, double bagged, and dumped at a landfill. The then-mayor of NYC called it "letting them go to sleep with nice dreams."

The next year, they donated the flesh to food banks. Did it make it easier to swallow?

I called the local 311 city service line to question the necessity of executing these birds and the lack of transparency in the decision-making process. I received this response from the then-mayor: "Dear Friend: Thank you for contacting me about Canada geese, and the measures we're taking to reduce the dangers they pose to aviation around our City. I'm glad you took the time to write . . ."

Why is it when a plane crashes into a bird, it is called a bird strike? How do we reduce the dangers to birds caused by plane strikes?

I was reading Norman Mailer's *The Armies of the Night* around that time, and was thinking about what Mailer calls "Negative Rites of Passage." Toward the end of the book, Mailer discusses a government spokesperson: "The spokesman was speaking in totalitarianese, which is to say, technologese, which is to say any language which succeeds in stripping itself of any moral content."

"There are negative rites of passage as well," Mailer continues. "Men learn in a negative rite to give up the best things they were born with, and forever. However much must a spokesman suffer in a negative rite to be able to learn to speak in such a way?"

A celebrated pilot safely and miraculously landed a plane on the Hudson River, when the engines hit migratory geese. This was used to justify this first round of killing. Matt Damon, playing a pilot on the TV series *30 Rock*, quips, "You know what a great pilot would have done? Not hit the birds."

This geese roundup during the summer molt turned into an annual ritual. The next year the radius expanded further into Prospect Park, where I walked every morning with Moo Cow. In the early hours of the day, as so many others were walking, biking, hiking, running, praying, playing, we spotted clumps of feathers, discarded zip ties, and a lake empty of geese. A few ducks went missing too.

On a day shortly after, I talked with another South Asian woman by the lake, who was there with her three children playing nearby. She came to the lake and visited the geese when she needed comfort. She tells me about her younger brother, who died in a motorcycle crash years ago, when she was pregnant.

"He was the light of my family," she says, carrying this loss as if it happened yesterday.

Her husband decided not to tell her the news until after her baby was born. Was she missing her brother or angry with her husband at that moment? Why did she entrust me, a stranger, to share her deepest pain, this lingering sorrow? Perhaps grief recognizes grief. We stared at the quiet lake together. What was she going to tell her children?

A writing professor friend of mine told me she lied to her daughter when they went for a bike ride around the park.

"Mommy, what happened to all the geese?" her daughter asked.

"They flew south for the summer," she told her.

What would the truth sound like? They were killed by the state.

≋

The community organized a vigil called "Hands Around the Lake."

A council member, who will one day become attorney general, said, "I was once an ugly duckling." When kids made fun of her at school, she came to the park and talked to the geese. "Though people may make fun of us and call us crazy," she says, "we know what we are doing here is right."

While walking out of the park after the vigil, the concerned parkgoers wondered what the ducks and the swans were feeling, with the geese noticeably gone. "I think that swan is in mourning," a woman who lived nearby said, pointing to a white, curved bird floating in the lake. "This week, he hasn't been the same."

The geese management policy evolved, and the city will opt for what they call egg addling. Oiling or shaking an egg to destroy or suffocate the embryo. The egg is still intact, so the mother goose continues to sit on it. Some feel it is the lesser of two evils. It's branded as a "nonlethal" measure. It sounds like psychological warfare to me. That poor mother goose, waiting. *Nonlethal* does not mean they've lost their cruelty.

My friend MaryBeth from Provincetown took me to see Mary Oliver at the 92NY. When Oliver reads her much-loved poem, "Wild Geese," so many people in the audience put their hands to their heart and sigh.

Why do I mention the geese? The birds who've come to call our city home. The ones who comfort the grieving. The ones who console a young girl who will grow up to be attorney general. The ones who, like Oliver said, "do not have to be good," but will inspire an entire auditorium to put their hands on their hearts.

They can be beloved and erased at the same time.

∼≋∼

What does a refuge feel like? On a hike in the Calanques with my fellow artists, we come across something called "the cave of immigrants," used by refugees into France in the nineteenth century. One by one, we go into this small entrance on a cliff. Sleeping and cooking quarters are carved out. But what I find most interesting is this rounded stone near the top of the entrance, which serves as a post to grab while climbing in. The stone is smooth, unlike the rough angular colluvium of the cliffs. How many refugees took hold of this rock to give it this texture? Erosion, here, feels like a welcoming hand, lifting you into the cave.

But what does it feel like to be unwelcome? Another afternoon, I am sitting outside my apartment overlooking the Mediterranean Sea, reading *Citizen* by Claudia Rankine. I am struck by her assemblage of ordinary racial microaggressions and their cumulative impact on bodies. Like the waves in front of me, memories suddenly come crashing to the shore. The mocking of my long name and the jokes about Red Dot specials at the local supermarket as a child. The relentless invisible labor I perform to make my ideas visible; the subtle ways my ambition has been checked. Each incident seemingly benign, but compounded like this, hit by one wave after another, I am reminded of this other way erosion works, how we are made small. It's not just my body and experiences. Systems, centuries of ideologies, have been devaluing so many kinds of bodies.

I think about the hairy plants hiding, and how sometimes being invisible is an act of survival. What would it feel like to be fully free in one's body? While I journal on the cliffs, I write down this duality: *Do not make a scene. You have to make a scene.* So much of activism is toggling between these two states.

Near the end of my residency, I visit the local swimming pool hoping to use the hammam and sauna. I want to sweat and relax my muscles before the long flight home. The door to the hammam is fogged up due to the steam. When I open it, I am startled to find four men in there, and shut the door. I did not feel safe entering a small, cramped space full of steam and men I do not know. I try the sauna. One man is lying down, occupying the entire bench. I am still uncomfortable but refuse to retreat. I enter and sit next to the door in case I need a swift escape. I notice ants on the floor. I pay close attention to them, so I don't step on them or vice versa.

Unlike the sauna man, I do not recline in some vulnerable position. I am hunched over with one hand on the door and the other covering my chest with a towel.

When did I become afraid of taking up too much space?

The sauna man asks if I am Italian. I say American. He is skeptical, looking at my brown skin.

"Américaine origine?" he asks. "Or Mexicaine?"

As a foreigner in a different country, I have been expecting the "Where are you from?" question. But so far, I haven't gotten the usual follow-up: "Where are you *really* from?" During this residency, when most people asked where I was from, I would say New York, and they would respond favorably with something like "Oh, I love New York." They accepted me belonging there without question.

The sauna man's inquiry was more familiar. I've had a lifetime of encounters essentially asking, "What are you?" I would answer, hoping it leads to understanding or acceptance, but more often, it makes me feel othered. I wonder if I should answer the sauna man now. *Do not make a scene.*

In distressing situations, the chimpanzees I've known made themselves really big, stood upright, and showed their teeth. Behaviorists classified these individual acts as threats. When you have three or more threats, it's called a display. *You have to make a scene.*

In the sauna, though, I am too tired to display. I concede and tell him my family is from India and satisfy his desire to classify my otherness. I stay the hairy plant, hiding under my towel, avoiding altercation.

―――

At a group dinner back at the residency, I ask one of the artists about the ingredients in the dressing he made. He explains it was oil, vinegar, garlic, shallots, and curry powder, and asks if I am allergic.

"Non, je suis végétalienne," I reply. *I'm vegan.*

"Oh, I forgot, I put beef in it," he says. "Pigs' blood," he adds, waiting for my reaction. He wants to upset or amuse me, and I don't want to give him the satisfaction of either. We live in a society where the torture and destruction of animals on a mass scale is so normalized that those who choose to question it are ridiculed.

*Do not make a scene*, an inner voice urges, knowing they will dismiss me if I do. I've become conditioned to be polite, to not express my rage, to show them I can take a joke. *You have to make a scene*, another voice pleads, because if I do not advocate for the animals, who will?

The next group dinner, I boycott. Guests are visiting the residency, and the other artists want to cook octopus. They arrange for a boat ride with the fishermen in the morning to collect beings in these threatened waters. I do not want to feign politeness or happiness about this meal of violence at a residency for artists exploring the relationship between homme and nature.

My friend Lisa and I decide to have our own dinner. Lisa is my kalyanamitra at this residency. She cares about animals like I do, and we have long walks, hikes, and meals together talking about coexistence, which will continue long after this residency is over, in New York and in Canada, where she lives. That night, we share a meal of lentils, grains, greens, fruit, and dark chocolate. Our absence at the octopus dinner is noted, but not understood. Was this making or not making a scene?

———

Perhaps the scene I have to make is on the page. I no longer want to be the hairy plant hiding in my body. I want to sprout unafraid in threatened environments and practice radical giving like the astragale de Marseille.

At the residency I review edits on my essay, "Are You Willing?," inspired by a line from Mary Oliver—"Here is a story / to break your heart. / Are you willing?" Oliver wants to break our hearts open, so they "never will be closed again / to the rest of world."

I write to practice a radical generosity. In my essay, I draw from my experiences editing the twenty-year-anniversary special issue of *Satya*. I wonder why hearts are reluctant to fully open. In her essay in *Cultural Anthropology*, "Witness: Humans, Animals, and the Politics of Becoming," ethnographer Naisargi Davé recounts a story of a woman telling her she could never volunteer at an animal shelter because she was "deathly afraid of caring too much." Davé then asks, "Is any other politics, I wonder, constrained by such a mortal fear of caring too much, of the heart bursting, the skin thinning, of not being able to rest again?"

In my interview with Davé in *Satya: The Long View*, I ask her to unpack this fear. She talks about concepts she calls the "descent into obligation" or the "tyranny of obligation." "The question becomes, once I have opened myself up beyond the ethical boundaries to which I had become accustomed, based on what criteria will I close the circle again? Once I care, how can I not care?"

Opting out of caring becomes a self-defense mechanism, a way to protect the heart. As a writer and editor, I wonder how understanding these fears can help reach the reader who reads with her hands over her eyes.

I listen to my friend Sunu on a podcast introduce the term *micro-inclusions*, a counter to microaggressions. Maybe the scene we have to make is one of inclusion.

This also reminds me of the Indian practice of kolam. This ritual Tamil women perform in the early morning, creating intricate mathematical designs out of rice flour, to feed ants, worms, and other species. This is a scene of micro-inclusion, radical giving, and losing cruelty.

# UNGOVERNABLE BODIES

I BEGIN TO NOTICE THE BODIES OF TREES. THE WAY THEIR toes bury themselves into the earth. The way their trunks bend toward the sun as if in prayer, and the way they extend their branches with offerings like the arms of a Hindu goddess. Wan and I have our own secret forest in the middle of Queens, where we take Mookie for walks. She is the one who teaches me how to pay attention. To the three trees guarding the entrance of the trail. To the one with the split trunk that looks like a peace sign. To the fallen branches she drags with her mouth, galloping at full speed.

I reflect on my twenties, devoted to vanishing primates in an orphaned forest. My thirties bore witness to how chickens and cows suffer and die, the massive scale of reproductive harms done to them. At forty, on the first day of a writing residency on the Mediterranean, I encountered that injured whale, and his dead mate.

When I return home from Cassis, Mookie is declining, and my fertility too. The next month, my body produces two eggs. They are retrieved on Saint Patrick's Day, and only one successfully fertilizes. The embryo makes it to the blastocyst stage, but on day seven instead of day five, which is more typical. It is tested and found to be chromosomally normal. "It's a miracle, at your age," my second reproductive endocrinologist tells me. Wan and I joke that this Seventh Day Blastocyst is a Seventh Day Adventist, because they are also vegan.

That spring, I see a statue of the three last northern white rhinos erected next to the Alamo Cube, near Cooper Union. A few days later, Sudan, the male, died. Il y a trois mais une est morte. His daughter and granddaughter, Najin and Fatu, may undergo IVF, like me. Do they want to? Do they understand what is being done to their bodies? Can we imagine a world where our children can thrive? Or are we all endlings, the word used to describe the last members of a species, or the last humans of a line?

Wan and I put the embryo transfer on pause. I want my body to recover from the stress of a retrieval cycle. The clinic's lab will close for a few weeks in the summer. I receive a writing fellowship for a week in Aspen. Again, Wan takes care of Mookie while I'm gone, and we decide to resume the transfer when I get back.

What I don't know about this invisible labor of fertility is how long it takes, how the process is subject to so many stops and starts, so many trials and errors. Years have gone by. I feel split between being in control of my body and not. Of being free to move in the world like I always have and being constrained to the tyranny of appointments, procedures, restrictions, superstitions.

In the interim we attend to Mookie's growing list of medical ailments. While I was in France, Wan outfitted a garden cart to wheel her in and out of our co-op building every four hours so she can relieve herself. We blended her food to make smoothies for her when it became too difficult for her to chew, gave her subcutaneous injections to keep her hydrated. We would hang a bag of fluids and watch it drip to form a hump of water on her back. Water is life.

We continue to be amazed by how Mookie learns to live in a body that is constantly changing, how she trusts us both with these new duties. After slurping her food smoothie, liquid spreads all over her face, and she lifts up her messy chin, and waits for us to wipe her clean. We work together to make these moments full of dignity, or dognity.

Our lives become circumscribed around hers. Wan and I no longer travel anywhere together, as Mookie needs one of us by her side always. There is the constant advocacy for her care, the questions to make the difficult choices, and the worry that despite all this care work, we are still failing.

When the timing finally aligned to transfer our seventh-day embryo, I took a couple days off work and read books on the floor with Mookie.

That summer in 2018, my body carried so much grief. On the fifteenth anniversary of my father's death, my childhood friend lost his father that same week. I rearranged my schedule to get bloodwork done en route to the funeral.

I cry throughout the services. Afterward, my doctor's office calls to let me know I'm not pregnant. That first long-awaited attempt at embryo transfer failed. Tears pour down as I drive back home to Wan and Mookie.

We have a vet appointment scheduled the next day, post-funeral. It is August. The vet says Mookie's body is shutting down, and that we should arrange for someone to put her to sleep within the next forty-eight hours so she won't suffer, and gives us some names for in-home euthanasia. It is a gift that the vet makes the timing of the decision for us, and that we can make it most comfortable for Mookie. That night, our family comes over to say goodbye. They surround her in the bed on the floor and the image looks like descendants gathered around a hospital bed of an ailing elder member. Before they leave, one by one they go into the room with her and whisper goodbyes. Our niece Lavanya gives a promise to Mookie, and sketches in her journal that night of a future business she will open called Mookie's Bagels, the O's being bagels.

On our last day with her, Wan and I do a final morning stroll in the sun with Moo Cow before the vet arrives. Then, we lay her out on blue wee-wee pads in the living room. The vet places her paw in foam, to make a print. Normally, the vet says she gives two doses. One injection to calm the animal, the other to terminate, she explains. Mookie is already so calm, she says that she doesn't need to give that first dose. I feel robbed of more time. We are ready and not ready. We put our heads on hers as she receives the final shot, lets out her last sigh, and urinates.

The vet says, "Take all the time you need," and steps out of the apartment. She comes back a few minutes later, which does not feel like all the time we need. We pick up her body and place it in her cart to bring her to the vet's van. She'll handle the cremation. Wan and I return back to an empty apartment that filled with her presence. *What do we do now?* We take a long walk in Forest Park, stopping at all of Mookie's favorite spots.

My friends Beth and Emily come over that week to sit shiva for Mookie, and our other friend Lisa joins virtually. I am grateful for them and our writing salon.

Wan and I decide to book a trip to Korea to visit Wan's parents, whom he hasn't been able to see in months. His mother is recovering from a stroke, and his dad's caregiving role resembles ours from the previous months with Mookie.

We have a twenty-four-hour layover in Honolulu where we swim briefly in the Pacific. We arrive in Wonju during a drought—the driest it has been in forty-six years. I download an air quality app to check on particulate matter. Wan's dad gives me a face mask to wear for morning jogs along Wonju Chun, where everyone is armored in full-body coverings to protect from the scorching heat and dust.

On August 15, Korean and Indian Independence Day, Mookie's ashes are supposed to be delivered to my mom's place because we will not be home to receive them, but the wires got crossed, and they are en route to our empty apartment. My cousin Hema goes to our apartment to wait for Mookie's ashes, but they never arrive, because the error was identified and the ashes were again rerouted. It is the middle of the night in Korea. I keep refreshing my phone screen, to track Mookie's ashes at various shipment facilities. I can't rest until she is at rest too.

Four pounds of what remains.

Later, I read Eileen Myles's book *Afterglow*, a pit bull memoir, and I return to the phrase Myles uses to talk about the beautiful and painful time of providing love and dignity to a loved one as they die: *radiant suffering*.

Jaya Aunty would say Work is God. Savi Aunty would say Work is Worship.

The closest thing I had to such devotion was taking care of Mookie.

What happens when you no longer pray?

Friends of ours, animal activists and goose advocates, get married on Labor Day at Woodstock Farm Sanctuary. I sent my regrets a few weeks earlier, not knowing Mookie's situation, but now we decide that we can go. On the way, we pick up Mookie's ashes from my mom.

Just like the trip Wan planned to Farm Sanctuary after my father died, this trip is a reminder of the things we care about. We are surrounded by fellow animal people who've cared for and endured so many animal losses. There is so much love and understanding in the room.

The wedding is beautiful. A flock of geese flies over the couple during the ceremony.

We attend a tour of the sanctuary. There are the cows: former dairy cows and rescued male calves from the veal industry. "Any time we put beings into this system of making money, their interests are not considered," the sanctuary director tells us. "When we eat animals, we separate families," she says. Earlier that summer, the news reports on the practice at the border of separating immigrant children from their parents and locking them up in cages.

At the sanctuary, we spend a lot of time with the chickens, "tiny, gracious dinosaurs."

"Mother hens make the best mothers," the director says. They listen for peeping in the eggs and turn them over.

After indulging in vegan wedding cake, I give up gluten and sugar. I try a new acupuncturist and travel to her in the early morning hours before work. I am so tired, I snore during my sessions. I prepare for another round of IVF. But there is no peeping in my eggs.

---

I am scared my body will fail me again, that modern medicine has been failing women, not believing women, not understanding women. We have been failing the planet. I believe those two failures to be related, the not believing women and the failing the planet. Flint, Michigan, is what happens when you don't believe women like Dr. Mona Hanna. I begin to lose patience for repeating myself so many times in meetings or appointments to be heard. I lose my tolerance for being interrupted.

There is an anger burning, a forest burning, and a pyre burning. I can't tell the difference between my grief and my rage.

A few months later, we stop trying, and Wan and I focus on each other. We do bucket list things. I sign up for my first half-marathon in April of 2019. Wan has been a longtime marathon runner, while I was the one who would run to catch buses. But I feel like if I put in the work, I would have something to show for it. After all the losses and failures, after so much laboring, I need a win.

I look forward to our long-run days, planning my distance to a vegan destination, where Wan also runs and meets me. The first route is about six miles to Champs Diner in Brooklyn. The last long run before the race is eleven miles to Rockaway Beach, and dining at Cuisine by Claudette. I enjoy seeing parts of the city this way and realize what is possible on foot. On race day in Flushing Meadows Park, I accidentally cross the finish line backward, because the course markers were taken down by the time I completed it. Wan and my friend Mikelle were there, and we celebrated afterward with vegan dim sum.

That summer, Wan and I plan for an epic trip to get through our pain and losses.

We decide on Seychelles and take scuba lessons in a Manhattan pool on weeknights to prepare. We learn to breathe, to regulate our ear pressure, to be buoyant. We dive and find Band-Aids at the bottom of the pool. We are famished after practice and stop by NY Suprema to get vegan slices on our way home, carrying our new fins on the train.

Why Seychelles? our friends would ask. Granitic islands in the Indian Ocean with the most beautiful beaches, mountains, and Aldabra tortoises! What we don't say is that we want to plan a trip to a place that is also free of Zika, a risk to pregnancy, and we haven't given up hope yet.

There is a haunting way that trying governs all decisions, even when you are not trying.

On our first hike in Mahé to Anse Major, the waves are temporary thieves. Wan loses his wallet in the water, and it returns to shore a few days later, just like so many honeymooners' wedding rings. The ocean seizes my fin, too, but then returns it. I watch it oscillate from water to land, stuck in between, just like my first language I thought was lost, but when I overhear a woman ask in Tamil what all mothers ask—*Saaptaacha?*—*Did you eat?*—I retrieve it. Maybe my Tamil was never lost, merely submerged.

Breathing underwater is like remembering this submerged language in this ocean that spans from India to Africa, like the mythical homeland of Kumari Kandam or Lemuria.

The first rule of scuba diving is *do not panic*. When I do, my shallow breath shoots my body upward. *Exhale all the way*, my scuba instructor signals to me. I blow everything out of my lungs and my body sinks. I inhale and wait for it to rise again. I want to be a seated Buddha, levitating in lotus position, in the middle of the ocean. I want my breath to let me safely hover over cities of coral. I want to transform into an elegant sea creature, like my father who always felt more at ease in water.

I am most scared sloshing around at the surface. It's the Southeast Monsoon season. Choppy, high waves. I need to calm my nerves to descend. The dive master says to me, "Just relax." And I hear Wan say, "Uh-oh." But I have to relax to survive. Underwater it becomes clear that panicking is not an option, I can only focus on breathing to survive. Uyir, *breath of life*, or as David Shulman said, the innermost core of a Tamil being.

On an excursion to Curieuse Island, we witness a giant free-roaming Aldabra tortoise yawn, and proceed to smooch another. A pair decides to nap under the same thatched roof where Wan and I seek shade. They approach us intently, dragging their shell luggage, step by step. Have they lost all memory of when humans harmed them, or does that also return?

These islands in the Indian Ocean are vulnerable to climate change and sea level rise, what the Tamil poets called katalkol, being swallowed by the sea. The seasons in the Seychelles are named after the wind direction. The southeast season is normally dry, but the rains that summer confuse the Seychelles black parrots into thinking it is mating season. We hear their mating call, but are told if they mate then, there will be no food for their babies when they are born. I check the news back home in New York City—heat waves, floods, and power outages. The footage from Brooklyn reminds me of Sittwe in Burma—wading in knee-deep water during high tide. El Niño bleached most of the coral in Seychelles, but on our dives we see a sea turtle who is old enough to remember how it was before. The damage is swift, the recovery too slow.

I am in awe of all the new plants we meet and their teachings. The black thorns on the palm trees in the Vallée de Mai serve as memory to a time when the entire island of Praslin was covered with free-roaming giant land tortoises, and the trees sought protection. The tortoises are gone, but these palms remember. Our guide reminds us to stay on the path as we look up at the gigantic thirty-five-pound nuts of the endemic coco de mer. "You are already in paradise, you don't want to get sent to the next." The trunk of this tree looks like circular discs stacked one on top of the other. You tell the age of the tree not by counting the rings outward, but by counting the rings upward. I think of pencil marks on a wall documenting a child's growth.

When I return back home to New York City, I yearn to return back to water. We take a trip with Gotham Whale in the Rockaways on our wedding anniversary in August. We see a humpback whale off the shores of Queens. She is logging—sleeping horizontally at the water surface. Whales are conscious breathers, which means that they only rest part of their brain, while the other part remains focused on staying alive. Part of me is focused on joy: my husband and I celebrating us, cleaner waters, and the increasing presence of whales choosing to migrate to our city. The other part remains in mourning, an activity these days/weeks/years that has come to feel as regular as breathing.

Soon after, the Amazon forests are on fire: How do we reckon with our complicity in planetary violence and forge an alternate path?

It is disorienting to be returning to fertility treatments at this time. Are we the Seychelles parrots, trying again at the wrong time? In September 2019, Wan and I try a new clinic that is focused on a more natural IVF that is more appropriate for women with diminished ovarian reserve. It's a slower approach, focused on quality and not quantity—retrieving the one egg I ovulate each month, creating embryos, and banking a few before we try implanting. There are still many decisions. Each step of this is precarious too.

Just when we are about to start egg retrieval, when the sun, moon, and ovaries have aligned, when all the hormones are optimal, and the follicles are round and a perfect 20 mm or so, we have to sign the waivers again. Initial here, sign there. Our embryos will be frozen and shipped to a lab, until they are used. What happens if Wan and I separate or one of us dies during the process? Who will get the embryos? What happens to the embryos if both of us die?

Legally, I know it makes sense to have a plan for such things. But in the moment, after years of the invisible losses, month after month, it is an added cruelty to be forced to imagine these other fates too. It feels like punishment. It feels like Heisenberg Uncertainty. It feels like Schrödinger's cat.

It feels like there is only possibility in uncertainty, and you lose it, if you try to know the truth. You always have to imagine more loss for any gain, more suffering, and there are infinite ways for a woman to suffer.

I lay in a reclined hospital bed, knees up, in a pink robe, with a blue hairnet and feet covers, like those provided at the Indian slaughterhouse. The room has a window in the wall, where the embryologist stands and confirms my name and DOB and that of my partner. She is cheery. The setup reminds me of the Airbnb we stayed at in Praslin in Seychelles, with a window from the kitchen to the outside dining area. Once the egg is retrieved, it will be passed to the embryologist through the window, like we passed our dishes to the kitchen sink.

I wake up in a chair lined with blue-and-white wee-wee pads, like the ones we used for Mookie. I hear the nurse ask the anesthesiologist if he's coming to their holiday party. I imagine this party of nurses, sonographers, receptionists, and doctors—the village it takes to conceive a child.

What no one prepares you for is all the decisions and uncertainty. Are the eggs mature, will they fertilize normally? Should you freeze on day three or push to blastocyst stage, where the embryo can be tested? But there is a risk the embryo will not survive to that stage or the testing. We cautiously freeze at day three, with the hope that the embryos will have better luck making it to blastocyst in the womb. All this invisible labor and decision-making to create eight cells.

It is November 2019, and when I read the news, I want to text Wan. I want to tell him there's a river of blood in Korea—swine flu, culled pigs, and rain. Wildfires burn koala bears in Australia. Bats fall from the sky when temperatures reach 107 degrees. Wild horses drowned in Hurricane Dorian, but three cows swept away are found on an island off North Carolina. Wan is biking to work, and I am afraid his wrist would buzz, and he'll take his hand off the handlebar while crossing the Triboro to receive the message and break my heart again.

This is the climate in which we are making embryos, and we don't know that a global pandemic is on its way.

~~~

*Overheard in Bodai Restaurant, Queens, New York, December 2019:*

"What a gift it is to have lunch with you," an older woman tells the young man sitting next to her. "I don't understand how grandparents move to Florida."

"I don't understand the appeal of Florida," her grandson replies, "except ruining elections. Could you imagine if Al Gore won in 2000?"

He pours her some tea. "What do you think about the climate crisis, Nana?"

"I think we can solve it," she says. "We have to."

"My generation is more pessimistic," he tells her.

"I don't blame you," she says. "But I've already lived through so much."

It is right before the new year, and I take the Q88 bus from work on Junction Boulevard to Main Street to return to my favorite vegan dim sum restaurant. The end of the year is a busy time for them, and the place is so packed that I share a communal table with this grandmother and her attentive grandson.

Sitting across from this pair, I watch how the two decide which lunch specials to order, and the care with which they debate and consider the other's perspective. It is a New York City pre-COVID-19, about to enter a presidential election year. In a few months, this restaurant will close its doors forever. I wonder what this grandmother and grandchild are saying to each other now.

Wan and I always celebrate the new year as Moo Year, the anniversary of bringing this joyous and loving pit bull dogster into our lives. So, we begin 2020 thinking about her, about us, and the uncertainties around caring for life in this world. I am feeling ready to have a dog in our lives again, but Wan isn't. He wants to wait until we live in a tiny house on a plot of land in the mountains, and not navigate dog life in our co-op in Queens again. But we both are public servants for New York City, and our tiny house fantasy seems years away.

Instead, we focused our efforts elsewhere. He on running, training for the Boston Marathon, and me on writing. There is also the unfinished fertility journey.

In January, my friend Lisa, my kalyanamitra from Cassis, invites me to an arts festival in Guelph where we discuss radiant friendships. I listen to her talk about trees. She thanks the trees for making the world. She offers us mushrooms, ramps, and nuts from the forests. She looks to trees to learn how to live with a precarious future. See how they produce in lean times, she commands. See how they give everything back to the earth when they die? She wonders, "How can we prepare for the uncertain future in ways that act tenderly with the present?"

In February, my friend Iselin texts about the unusual warmth of the sun on her body. We message about finding delight to anchor us when much is destabilizing. We write to each other about grief, the loss of a parent, our beloved animals, endlings. We look to find language to describe these times. Whether our bodies can bring children into this world, and what kind of world it would be. The pendulum is always swinging. Can we grab it for a moment, bring it back to center, and hold still for a moment?

Wan and I visit his sister and our niece in California in February, before attempting embryo transfer again. The Joshua trees looked lonely, like windmills facing the wrong direction. The tumbleweeds' billowy frolics betray their thorny nature. The neighboring valleys and ranges are named after death and devils. In the harshness of the wind, sun, and heat, the orange poppies give life to the parched desert hills.

We are worried about how a new virus could impact Wan's family in Korea. We get brief updates from daily KakaoTalk messages. *Day 1278*. Abeoji sends a Kakao message counting the days since his wife's stroke. My husband and his sisters reply *yes*, in Korean, acknowledging their father. Eomeoni recently broke her hip and Abeoji visits her in the hospital daily. Now, visitors are barred to prevent the spread of the virus. Abeoji goes to the hospital only to be refused entry, with no information on how his wife is doing. We don't know if Eomeoni understands where she is or that it is her husband's birthday. The CDC then urges US Nationals to avoid travel to South Korea. What is the untold emotional toll of being isolated from your family? Love in the time of corona.

Wan and I take a distanced, masked walk in Forest Park early on in the pandemic. Even the trees reach for each other. We see a hawk, low to the ground, struggling to fly. I text my animal friends David and Beth asking about pandemic protocols for transporting injured animals to Wild Bird Fund, our local wildlife rehabber. Before they respond, he flies away. "I hope he's okay," I say. The hardest part is the not knowing. Not knowing when Wan can visit his parents in Korea again. Not knowing how India can handle something like this. My gastroenterologist writes me a note: "We currently do not know what effect your medication will have on the body's response to COVID-19." My heart is heavy for our borough of Queens. We do not know when we will have enough hospital beds or ventilators. My fertility clinic cancels my upcoming transfers. There are risks about COVID and pregnancy. There are unknowns about COVID and birth. They do not want to crowd hospitals with non-COVID matters. New York City is under lockdown. Could there be a clearer sign that trying to have a baby at this time is not the best idea?

I am relieved to put a pause on trying during this time, but I do not know when or if it will be possible to try again. We decide to apply for a rescue dog, since we will now be home all the time for who knows how long.

Asta grounds us in the present. Her name means *divine strength* in Greek, or *bright as a star* in Latin, or *smile* in Hindi, but we pronounce it like *Hasta* in Spanish. *Until.* I say *Asta Mañana*, kissing her good night. *Asta la vista*, when I leave for a few minutes to do laundry in the basement. Amid a pandemic, we perpetually wait "until." Asta, too, has been waiting. She was shot in the face and dumped over a fence at a shelter in Georgia. The shelter staff found her in a flowerpot the next morning. She was transported through an invisible animal rescue network from Georgia to New York and waited almost two years for a home. Asta Ahora. This moment is unlike anything we have ever experienced, and we stumble toward new language and rituals together.

Crematoriums in New York City operate twenty-four hours a day. Quick-burning caskets are recommended. New York invites out-of-state funeral directors to assist. Hart Island prepares for interim burials. Seat belt webbing is repurposed as handles for body bags—material intended to save lives will instead carry the dead. The first death at my agency is announced—our mailroom supervisor. The messages we send and receive begin with *Hope you are well*, and end with *Stay well*—by which we mean alive. We answer *Okay* to *How are you?* by which we mean alive, but not okay.

Asta cuddles up into a number 6 and lays next to me while I work from home. She rolls over when she wants her belly rubbed and puts her head under my hand, demanding more if I stop. She doesn't respond to the construction noises of the elevator repair in our building or the seven p.m. clapping for essential workers. But she lives in fear of her own species.

We have extreme social distance dog walks. It's like a game of Pac-Man, and we keep turning down different streets or park trails to find different ways back home. We are supposed to redirect when she reacts to another dog. Mark and move. We carry a squeezable tube of peanut butter and try to associate dogs with good.

Wan and I begin calling each other and wearing headphones during our Asta walks. I hold Asta and Wan will be twenty feet ahead on lookout. It's like Asta has her own secret service detail. I begin a conversation, and Wan listens, but responds with other information: a warning of an oncoming dog, a pile of horse poop, or broken glass. My story resumes or is paused for a later time. Wan may start his own story. I listen, but then will respond with only updates on Asta's bowel movements. It's happening! Good girl. Who did good?

We all walk in hypervigilance.

*Listen.* Asta's ears go up when she hears the jangling of keys, the slamming of car doors.

*Look.* Her eyes look up at the squirrel tussle on the telephone line.

*Leave it.* We direct her away from the lingering leftovers on the curb.

When we go for headlamp walks into a dark forest, we see the yellow glow of a family of raccoon eyes perched in a tree. The non-human primates of North America, or the descendents of those babies Mookie found in a diaper box years prior.

Asta's head peers off to the side like Norman Rockwell's *Triple Self-Portrait*. She must confirm:

*Human or Dog?*

*Dog or Bag?*

*Bag or Cat?*

*Cat or Raccoon?*

Our job is to get her out of fight-or-flight mode and back into pack mode. It is what she does for us, curling up next to us requesting affection, redirecting us from our anxiety screens, back to pack mode.

I used to run with headphones. Now, I pay attention to the birds. Before, I couldn't match their feathers with a name or their names with a song, so I googled *Shazam for Birds*. The pileated woodpecker sounds like my cousin's laughter. The rock pigeons with their spooky hoot are perched under the elevated tracks of the Long Island Railroad. The gorgeous cardinal, in my mother's favorite red, whistles as he struts up Park Lane South. The ubiquitous orange-vested robin cheers *hello, how are you?* Within a pandemic, I am finally getting to know the neighbors.

---

The dead wait in hospital morgues, in makeshift burials, in refrigerated trucks, in unrefrigerated trucks. Ooze dripping from the trucks outside the funeral home on Utica Avenue seeps into the sewer system. The body bags are stacked in U-Hauls, like a college student's possessions hastily collected as they fled campus. The dead wait at BCPs—Body Collection Points—to be claimed, to be transported, to be buried, to be cremated, to be registered in a database called eVital. The newest BCPs are on Brooklyn's 39th Street Pier. The view of the Statue of Liberty is no substitute for dignity.

Asta helps us keep track of time. The day is structured around her walks: chatty bird hour in the morning, the sun peering through leaves; quick pee before lunch, afternoon stroll down the block; nighttime headlamp hike under the railroad tracks. Each month is marked by her medication. Each week, we learn—nose-touch cues and stealth U-turns; walking side by side; waiting at corners—skills for distancing outside. Inside, white noise dampens the jolts of city sounds. A roaring river runs through our apartment while Asta naps and we work.

We see images of police batons and pepper spray. Bodies vulnerable to a virus and a state. Pro tips like Maalox and water; the phone number for the National Lawyers Guild written on your hand. Lessons learned for lessons not learned. If we say his name, George Floyd, we honor his Momma who said it first, whom he called out for last. If we say her name, Breonna Taylor, we can hear Bre, her nickname, or say Breezy because her aunt thought she was "a cool cat." If we say his name, Elijah McClain, may we hear his music—violin concerts he played for shelter animals to calm their anxiety. If we say their names, Amadou Diallo, Eric Garner, Sean Bell, we can say their last words: "Mom, I'm going to college," "I can't breathe," "I love you, too."

~

The alarm is now set for 4:30 a.m. to get out before the other dogs. We spend an hour in Forest Park with headlamps, face masks, and Asta loose-leash walking by our side. We avoid her stress triggers. We only see a few dedicated runners at that hour. One morning we see an opossum and the tree blowdown from Tropical Storm Isaias. We learn about stress stacking in dogs. Cortisol levels can linger for seventy-two hours, bringing one closer to panic threshold. We manage the environment instead.

🐕

Our trainer suggests a behavioral veterinarian for Asta. We start a medication regimen to help Asta stay below threshold. The goal would be that either her reaction intensity, frequency, or reaction radius decreases. We are told it may take several months to find the right medicine or dose to work. It feels familiar.

🐕

During the pandemic, a phlebotomist comes to my home to draw blood, so I can avoid going to a clinic. The doctors are checking my inflammation levels before resuming fertility treatments again. I tell the phlebotomist that my veins are tricky, and he should try my forearm or hand. Having had regular blood draws for the past several years, I have good intel on my own veins, and usually only the more skilled technicians can do it in one shot. He is confident he can get the vein in the bend in my arm. Of course, he can't, and has to try again where I initially suggested. I get lightheaded. He tells me it is "mostly psychological," and to lower my mask to breathe. I am furious. Why is it that I am not heard or listened to the first time? I have to do so much extra work to get my point across. I put this much work into *everything*.

My bloodwork shows that I have elevated inflammation in my body. Would it be different if these experiences were an anomaly and not the norm?

My doctors' offices reopen again by summer when the COVID levels go down in NYC. My GI doctor says it would be good to get pregnant before a second wave in the fall. As if I have any control over any of this.

To pursue bearing children in a time of COVID and ecological collapse feels like something heavy is always on my chest. It is August and I am thinking of our trip to Burma, seven years prior. We visited a cheroot factory in Inle and Wan bought a small case. We don't smoke, but back then we could allow ourselves the fantasy of sharing our anticipated good news with Burmese cigars. Over time, we lose the cheroots. Is it a sign? But in August of 2020, seven years later, I find them. Is it a sign? The next day, I learn my next frozen embryo transfer would be on my birthday. I rekindle excitement for the possibility of cigars.

# UNGOVERNABLE BODIES

~~~

I am on the way to my transfer and receive a text from my mother on my birthday telling me about the day I was born. I have not seen my mother in many months, since the lockdown began. I feel like there are three generations connected in this moment. I am forty-three and learning my birth story for the first time and hoping to share it with this day-three eight-cell embryo.

~~~

I learn my transfer doesn't work in the most inappropriate way. "I'm sorry, Ms. Iyer, you aren't preggo," my reproductive endocrinologist says. "Excuse me?" I say. The pregnancy test was negative. All the stories I told myself to think this time would be different suddenly feel foolish—the cigars from Inle, the coincidence of a birthday transfer. All the doubts resurface. Does it make sense to bring life into this world?

This disappointing news comes as anxiety rises about the potential return to school in New York City, where Wan teaches. Then, a disease of the gut topples a head of state. "Ulcerative colitis is finally getting its due," Wan jokes with me. News article explainers about colitis emerge when the head of Japan steps down to focus on his health. UC is an inheritance for me. I have my father's fluttering gut. A group of butterflies is a kaleidoscope, a group of caterpillars is an army. UC is a body at war with itself.

~~~

A five-letter word is my longtime companion. I learn new variants: *ambiguous grief*: Like the high school and college seniors missing out on their graduations and proms during a pandemic. *Anticipatory grief*, waiting for an imminent loss, which may or may not come. Both on personal and ecological scales. *Disenfranchised grief.* An invisible pain not recognized or publicly ritualized. The grief of all the invisible labor, of trying to reclaim space in a society trying to erase you. There is insufficient language about grief. There are losses that are so private and misunderstood. I want more language to capture the layers of loss on planetary and personal scale, that recognizes the love and resiliency in mourning. G-R-I-E-F sounds like a sneeze, something expelled from the body. But I need to cradle it. I want a word that lingers like a lullaby, something you learn to carry in your body.

———

The West Coast is on fire again, and I am heartsick for this place where I once lived. The fires have blotted out the sun. Orange is the color of wisdom, my father would say. I look for wisdom in that tangerine sky. I can only remember the bluest hues, sitting outside Vik's Chaat house eating dosa in Berkeley, walking around Lake Merritt in Oakland, and the views from San Francisco's peaks, where all the water tanks are located. I watch a drill rig hammer drop and descend and return with soil samples for me to log. I look up only to look down. My roommate from grad school texts me: "Today the sky is brown. San Francisco so much worse. I don't know how things will get better. It's just getting worse every year." Wisdom does not bring comfort.

———

We talk about major storms or earthquakes by their frequency; a one-hundred-year storm or one-thousand-year return period. The larger the magnitude, the less frequent the event. If you lived through the devastation, it was unlikely you would experience it again in your lifetime. We are now in a period of high magnitude events with short return periods. A hundred-year storm that occurs every few years; a hurricane season where the letters of the alphabet are exhausted. It's not just storms bringing the frequent devastation. Fire, zoonotic diseases, habitat loss, inequity. Each week brings more pain than we thought we could hold. One week began with the news of forced hysterectomies of immigrant women at detention facilities and ended with the passing of Supreme Court Justice Ruth Bader Ginsburg.

―――

I submit to an endometrial receptivity analysis—a mock transfer to check about the timing of the transfer. It is an optional procedure, an IVF add-on. I wanted to do it before my birthday transfer, but I grew impatient; COVID delayed this already by several months, and I was worried about a second wave and having only this limited window. I should have done the ERA first. I learn that I am "early receptive," and the timing of my transfer should be shifted twelve hours.

―――

I have two embryos left and only one transfer left that is covered by my insurance. We transfer them both. Twins are risky but even one pregnancy is still unlikely. I always thought I would have twins because Patti had two sets among the thirteen children, and they say it skips a generation. For some reason, as a child, I think this will be my fate. I loved learning about twin language, about deep bonds and understandings. I want this for these last two embryos. This is our last shot. During the transfer, "Like a Girl" by Lizzo and "Brown Skin Girl" by Beyoncé play on repeat as I lay in stirrups with a pink robe, face mask, and blue hairnet and feet coverings. When the doctor arrives, I video call Wan and Asta. On the display monitor, we see our compacting cells get launched into my womb. It's like watching a moon landing. They tilt the table after, and I am suspended upside down like an astronaut.

―――

I take the next two weeks off of work and focus on writing. I attend a virtual writing retreat and am asked to record my dreams. I dream of the stress of caring for sea creatures, of not knowing how to answer the question, "Do you have children?" I write in my journal, "Even in my dreams, I am mourning."

---

I find hope in a tarot workshop with Rahna Reiko Rizzuto. There is a light in the womb. The tarot card drawn as part of a writing workshop is the Nine of Birds, to represent the urgency in the work. It's a card of grief, of emerging from loss. What wisdom can the owl share? What will you create from this darkness? There is a light in the womb. I meditate on that light.

---

The Monday after the workshop I get blood drawn for a pregnancy test. When the phone rings a few hours later, I am nervous. I am expecting a call from the doctor's office, and instead it is our behavioral vet calling. She says that many dogs who are reactive have allergies and suspects that allergies are also the cause of Asta's anal gland issues. We ran an allergy panel earlier in the month. "Her results are very impressive," she says. Asta was tested for almost a hundred different allergens, and she is allergic to all but three—feathers, mosquitos, and grain smut mix, a type of fungi. The vet thinks Asta's irritability and reaction to other dogs could be intensified, if her allergies are already making her feel crummy. We make a plan to develop an immunotherapy vaccine for her, which will take several months to work.

The allergy results are information. There is a solution. It is hopeful. Asta and I are both dealing with ungovernable bodies. Wan buys us two multirow pill boxes—one for me to keep track of fertility supplements, and one for Asta for her daily medications. I think it is only a matter of time before we mix up our regimens, but it hasn't happened yet.

The second time the phone rings that morning, I am bracing myself for the same sad news I have heard twice before. But this time it is different. The result is positive. The doctor says the words, "Congratulations, Ms. Iyer, you are pregnant." There is a light in the womb. But he is cautious. My HCG is not high enough to be reassuring, so I return two days later.

My mother calls. It is too early to be sharing this news with anyone, but I tell her because I don't know if or when I'll be able to say this to her. I'm pregnant.

On Wednesday, my HCG increases but does not double. That is not a good sign. Again, there is pessimism from the doctor, but he tells me to return on Friday. I get my bloodwork in the morning and return to remote work from my two-week vacation. The doctor's office calls and my HCGs have doubled. Good news. I return the following Monday and they more than double and I am told to come back that Saturday for my first ultrasound.

~~~

That week is the 2020 election in the United States. The votes haven't all been counted, and the entire country is in this place of uncertainty with me. I do not want to pre-grieve until I have all the information. I read about a pod of over a hundred short-finned pilot whales that are stranded on a beach in Sri Lanka. I look at the photos of residents, young and old, wearing masks to come witness this event. Despite the global pandemic, they lend all the hands they can. They spend all day and night working with local rescuers and the military to push the whales back to sea.

~~~

On Saturday, my ultrasound is scheduled at a later hour to make sure one of the doctors will be there. When I arrive in the room, a sonographer is there. I have no problem with her doing the scan but I was told to come at this time so a doctor can do it. She explains that they just want a doctor available in the office "in case, God forbid, anything goes wrong." She asks if I prefer to wait for the doctor to be available. I am used to her doing my previous scans and do not object. She thinks she sees something, but is not sure, and runs to get a doctor. I wait, for "God forbid anything" to go wrong. The doctor arrives. He does not see anything. He thinks it's a failed pregnancy and worries it is ectopic, but I have to wait for the bloodwork. I don't understand how it could be ectopic. I saw you place the embryos in the uterus, I tell him. He says they could still migrate to the fallopian tubes, that ectopic rates are even higher in IVF pregnancies. Did I miss that fine print on all the waivers I signed? I leave in tears. The sonographer offers me water and tissues on my way out. I'm on my way to my infusion at Mount Sinai for my ulcerative colitis treatment. I have an hour to wait, so I order an avocado toast from a café across from the hospital and sit outside to eat, lowering my mask. The doctor calls with my blood results and my HCG increased from 300 to 1,128. "You are still in the game," he says. There's still a light in the womb.

I have a hard time hearing the doctor because there is so much honking and cheering in the streets. People have their phones out and are video recording. I look around and do not see any spectacle. I do not know what is happening. I check my phone and realize the election has been called. I scream with hope and relief in the streets with everyone else.

---

I return for imaging the following Monday, and there is a smudge on the screen, but the doctor is not certain. This is the other doctor in the practice. She is more hopeful, and not ready to call it yet until we have all the information. She orders more bloodwork to check for inflammation, other indicators of potential ectopic. She sends me to a fancier imaging place to see for sure. During this period, this practice has been doing very well. They move from a lower floor to a higher floor with more space. They have a grand piano in the lobby. They hire a pianist to perform on certain mornings. I do not understand why they don't invest in the better imaging themselves, instead of sending me out to another place.

I go there the next day, and they find a gestational sac in the uterus. It is not ectopic, out of place. They take a 3D image to confirm. I wonder about the second embryo. They search the tubes and it's not there, and confirm just one intrauterine pregnancy, or IUP as the report says. The other was likely adsorbed, or absorbed? Does that mean he/she/they are a part of me now?

---

I return the following week. The sac has grown from about 3 mm to 5 mm, but no yolk sac can be seen yet. My HCGs rise but do not double. I return again the following week. The sac increases from 5 to 10 mm, but still no yolk sac can be seen. The IUP is confirmed, but the viability still inconclusive. This is the language on the paper I leave with. Each time, Wan and Asta drive me to the appointment and wait in the car. We stop by Randall's Island afterward to let Asta run around in empty ball fields.

---

I return the following week and the sac stops growing, possibly shrinks, and still nothing is visible inside. My doctor now thinks it is safe to call it. I have options to remove the pregnancy—chemically, surgically, or naturally. I am grateful there are safe and legal options for me, but do not want to make this decision while grieving. I call one of my bestest friends who has been on this roller-coaster ride with me. I opt for natural and continue to wait.

※

During this liminal time, when I am still pregnant and soon to be not pregnant, I am invited to speak at an Extinction Rebellion literary event, On the Brink, for Lost Species Day, with Margaret Atwood, Amitav Ghosh, and so many other wonderful writers. We are all still on lockdown, on our computer screens spread across continents and time zones. We each are given five minutes to tell a story of one endangered or threatened species. I talk about chimpanzees. I share stories about caring for my orphaned small smalls, who lost their mothers, in a forest that lost its elders. I recount Washoe's story of infant loss and her empathy in recognizing this loss across species.

※

One week later, I start bleeding. I have trouble naming it, because I hate the word *miscarriage*, as I hate the word *infertility*. How cruel to be defined by these terms and blamed for them.

I read this beautiful yellow booklet written about Jaya Aunty, *Tulasi Mata. Mother of Tulasi*, her towering plant. "Far too many people have said Jaya Aunty has so much time for others because she doesn't have kids. What's worse is that they'll say she never had babies of her own," her friend Niki writes. Jaya Aunty was married at twenty-five and got pregnant for the first time at thirty-seven. "They were beautiful babies," G. S. Uncle says of his twins, who are born in 1977 and die shortly after. I grieve for my cousins who share my birth year.

"Once it was known that the kids were no more, Aunty was lying there on the hospital bed, helpless. Till today, she remains grateful to her mother-in-law who gave her a tall, warm glass of Horlicks when everyone else left after giving her their dose of rawkish barbs."

I do not know what exactly she means by *rawkish barbs*, but I have some sense of it. Some form of superstition or interpretation of karma, some way to explain the unexplainable, that helps others deal with their discomfort, but offers no comfort to the bereaved.

"Now when I think of those days, I feel the pain but when it actually happened, I never understood what they meant," Jaya Aunty tells Niki.

Perhaps I have experienced my own dose of rawkish barbs. The "What about this? Have you tried that?" or the "At least, it's not this or that." Words that make you feel like a failure and diminish your pain. What you really want is someone to see you. To know that you didn't fail, but that you were failed.

I find a red blanket Jaya Aunty knitted for me after I was born. I think about our trip to Bodh Gaya, after my father passed away. I learned then that women are allowed to perform the death rites in Gaya, but only after my brother had already done them. Jaya Aunty saw my disappointment and asked me to do hers. It doesn't have to be Gaya. "You can do it anywhere," she tells me. I agree.

I do not know all the hope or pain she held as she knitted this red blanket, but I wrap myself around it as I grieve.

~~~

This moment I transition from pregnancy to nonpregnancy lasts three weeks, and is filled with other losses.

There is no singular grief. Dinesh Uncle was ready to go to Savi Aunty. Savi was ready to go to Adi. Adi ready to go to Patti. Patti to Thatha. What is the word for losing generations of elders?

And what is this other invisible loss? The first ultrasound looked like *Starry Night* to me. In the next one, I knew what to look for—the anechoic chamber, the dark animal eye, my tiny gestational sac. I waited and waited, until there was no more light in the womb.

CRY PLEASE PERSON HUG.

# BEGIN AGAIN

**M**Y DEAR READER

Did you eat?

Have you had enough water?

When was the last time you laughed so hard you cried or cried so hard you laughed?

Did the sea swallow the land, and did we retreat inland like the Sangam poets?

Did we return to water, our first home?

What happened to the trees? Did we save them, or did they save us?

You survived, though.

Tell me, how did it end?

How do we begin again?

# SPEAKER OF RIVERS

THE INVISIBLE RIVER IN MY INVISIBLE CITY IS FILLED WITH invisible losses, invisible grief, invisible violence, invisible kindnesses, invisible labors, invisible infrastructure, invisible poets, invisible homelands, invisible histories, invisible archives, invisible ancestors, invisible descendants.

What is it I have come to say?
I see you.
What is it I have come to ask?
How can we be without sorrow?

The final card in Reiko's tarot writing workshop was the Speaker of Rivers. The mouth of the river is not where she begins but where she ends. What has she come to say? What has this journey taught us? The river reminds you to be a storyteller and that your story is not just for you.

# COLLABORATORY

I travel to speak at a conference in Guelph connecting scientists, poets, and artists engaging with environmental issues.

Madhur invited us to this *collaboratory*, a collaboration + laboratory.

It feels like a sangam, like a confluence of rivers, or the gathering of Tamil poets.

Catherine wants science rooted in tenderness. Can we re-imagine disciplines as zones of care?

Amy, who writes physics poems on subatomic and cosmological scales, thanks me for letting her know there are vegan muffins. Veganism feels like the final frontier at environmental conferences, she says. We chat about the chimp Washoe's ASL neologisms.

Forrest wonders, when the language of facts and information has been insufficient to move us, whether poetry can re-sensitize us?

Karen speaks of the importance of trialogue—three ways of knowing. Bring together peculiar capacities—working with felt, translating Japanese, and managing New York City's water supply.

Madhur talks about weddings in a mosaic landscape in India where there are patches of forest in the grassland. The Indigenous group who lives there only has a population of one thousand. They need both forest and grasslands to have a marriage. The guests must sit in the grassland, while the couple must stand in the shade of the forest.

Natasha's talk is about foxes (polymaths) and hedgehogs (specialists).

Rae begins a poem when something puzzles her. Why does this music make me sad?

Karen says every poet is used to being the dessert at a science talk.

Madhur says there's no dessert here. Everything is a main course. Eat your vegetables.

***

I walk through my catena of governing questions:

How do we plan and build infrastructure to avoid environmental impact?

Can we acquire knowledge without violence?

How do we repair the harms we have done to other species?

Can we avoid exacerbating multiple compounding crises?

How do we disentangle ourselves from systems of harm?

Can we sit with the scale of personal and planetary sorrow and emerge with truth and compassion as our guides, and not despair?

# BAS BAS BAS REDUX

Enough. When will it be enough? More than ten years have passed since I visited those chicken factories in India. I continue to pay attention to the various outbreaks of avian and swine flu. Ebola. And then this other zoonotic disease jumped species and travels the world, killing millions. During this pandemic, global travel came to a halt. I dreamed of India, and did not know when I will be able to go back again. Ganga filled with corpses. I quarantined in our apartment, leaving only for dog walks.

The American Veterinary Medical Association (AVMA) defines depopulation as the "rapid destruction of a population of animals in response to urgent circumstances." The first time I heard the word is after the first incidence of mad cow disease in the United States. In Europe, I read, they set the cows on fire. The word resurfaced throughout the years—and I learn of a center to assist with "depopulation and disposal" during COVID-19. AVMA states that "a depopulation plan must be informed by disposal." Destroying bodies is coupled with disappearing them. Words that convey cruelty disappear too.

They say *cull* when they mean *kill*.

They say *depopulate* when they mean *eliminate*.

They say *Ventilator Shutdown* when they remove the air supply.

They say *Ventilator Shutdown plus* (*VSD+*) when they add heat to roast them alive, or add carbon dioxide to speed the process.

They develop these techniques in the classroom and the laboratory.

They get others to do it in the field.

When a Colorado inmate tests positive for avian flu, it means they use prison labor.

They graph "Time to Silent" to document how many minutes it takes until the animals stop screaming.

They say *silent*, when they mean *dead*.

Bas. Bas. Bas.

I think about the birds—caged and free. What do they think when this latest strain of avian influenza takes over their bodies? The news does not report on their feelings, process their suffering, wonder how they explain what is happening to their bodies and to their flocks. The news does not consider what they love about their bodies despite what men have done to them. My friend's sanctuary is hit, and from her I learn about their symptoms: the purple-ing of their flappy wattles, combs, and legs; their disorientation; their misshapen eggs; the swelling of their heads; and their coughs, sneezes, and runny noses. I read of the sudden death of seabirds, and how shorelines have become that liminal space between dwelling and dying.

I wonder how the birds situate their own pain among their collective suffering—both boundless.

Bas. Bas. Bas.

# SHIZENGAKU

At the mouth of the Tavoy, the river spills all that she is, knows, has seen, and carried into the Andaman Sea. What secrets does she whisper? Maybe she reveals when and how the swimmers—the long-tailed macaques—arrive from distant lands, swimming through her waters to the forested rock. Does she remember when the square masonry tower with a round lantern appeared, flashing white every five seconds? Or my Patti birthing the twins—Amirthalingam Uncle and Sachi Aunty—on this lighthouse island?

Two years after Thatha and Patti left Mibya Kyun, *light island*, and Burma for good, a British scientist, H. C. Smith, came around collecting monkey specimens—skins and skulls. The date of capture is listed as 23 April 1936 in the *Systemic Review of Southeast Asian Longtail Macaques* published by the Chicago Field Museum of Natural History.

Does the river carry news about the current macaques on Mibya Kyun and their coconut coup to their family in other lands? Do the Mibya Kyun monkeys now know that, oceans away, their kin are stolen for laboratory experiments, and how their demand has increased since the global pandemic? Do the seas learn about the hundred macaques from Mauritius who flew to JFK and were loaded into a truck that crashed in Pennsylvania on their way to a medical research facility? Three monkeys escape and are re-captured. The same three monkeys are said to be humanely euthanized in the same article that says they are killed by gunshot.

Do the waters warn the lighthouse monkeys of the latest collectors of their bodies?

I have been thinking about these macaques from Tavoy, and reach out to Yuzuru Hamada, a primatologist in Japan. He writes, "Myanmar long-tails have splendid way of life." He tells me how the macaques collect shells and smash them open with stone or hard materials. "Even they extend this subsistence behavior to crack-open hard nuts (*Terminalia catapa*) which are often found in sea-shore. This traditional behavior is transmitted generation into generation, and also some migrating individuals transmit from population to population, perhaps swimming across sea to other land-mass."

I told him about the long-tailed macaques on the lighthouse island, and he wrote, "it is natural that long-tails prevailed at the lighthouse in some islands in Andaman Sea, as they know the way of navigation."

There was something about his response that charmed me, how it was more poetic than scientific.

Years later, I read Takayoshi Kano's book about bonobos, *The Last Ape*, which has been translated into English from Japanese. The translator notes, "the reader may be struck by the use of anthropomorphic language, a characteristic of traditional Japanese primatological work."

I've always felt anthropomorphism—attributing human characteristics and emotions to nonhuman beings—was based on a false assumption that these characteristics were solely human to begin with (how anthropocentric!). I wanted to know what

Japanese primatological work was and found a lineage tied to Kinji Imanishi, who acknowledged nonhuman primates have culture.

One September day in 1953 on Koshima Island, a schoolteacher named Satsue Mito saw a one-and-a-half-year-old female monkey washing a sweet potato in a freshwater stream. The potatoes were provisions given to the wild monkeys to acclimate them to their newest human observers: Kinji Imanishi and his research students.

The researchers gave this monkey the name Imo, *sweet potato*. The primatologists were trying to document the three components of a cultural phenomenon: "emergence, transmission, and modification."

Soon other monkeys would take their potatoes to the ocean, to not only wash, but to add salt to their food. Some used water to cool the provision of a hot potato. Seagulls, too, waited on the shores, to salvage whatever the monkeys left behind.

Tracing a string of acknowledgments in the library, I found this wonderful little book by Kinji Imanishi—*Seibutsu no Sekai (The World of Living Things)*.

Published first in 1941, Imanishi writes this short book quickly and with urgency, because the war had come, and he could be called in to service or die or both. He wanted to leave behind not a scientific treatise, but "a self-portrait," to provide a record of how he came to view the world and all the beings within it. He writes about dabbling in entomology, ecology, and observing gorillas and chimpanzees.

"I have done quite a few things, but it seems to me that I have been consistent in working on the problem of 'What is nature?'" he writes. "And I feel that it is not the constituent nature as represented by such-and-such-ology, but total nature, that I have been in constant quest of. What I have been seeking all this time is shizengaku."

I practice holding this new word in my mouth.

"Shizengaku does not fit within the general scheme of academic disciplines," Imanishi writes. "It seems that, though the word nature is used more than ever before, there has never been a time in history when people have had such a small realization of what nature really is."

He goes on to say, "We must teach [students] that nature is not matter, it is [a] living thing; it is the colossal maternal body, the giant, the behemoth within which we, along with all the other myriad creatures, have always been nourished."

His first chapter was on similarities and differences:

"If we are concerned only with differences, we will find everything is different; despite this, is it not wonderful to learn that nothing exists in complete isolation or does not have something similar to it?"

I stumble upon this next line, which reads like my writing philosophy, making links and seeing connections between everything.

"Although the things of this world are various, there is nothing with an absolutely independent and unique existence. In this wide world everything has something else similar to itself."

Imanishi looked at animals as members of societies with cultures. I think about Yuzuru Hamada, and his message to me about long-tailed macaques on Mibya Kyun. He, too, focused on their culture, which seems like a lineage shaped by Imanishi.

"I would say that living things have a wholeness and autonomy," Imanishi writes. This contrasts with the British scientist who saw the lighthouse monkeys as merely skins and skulls.

I recall how catenas are described in soil science, a series of distinct but connected soil layers. Imanishi introduces the concept of *affinity*: "Affinity refers both to the relationship of blood and of soil. It is the relationship of closeness and distance in historical development and in society."

I have an affinity for his thinking, the way he recognizes other knowledges that biology is too limited to perceive: "Therefore, it is not at all surprising that shamans and poets could talk and listen to nonhuman life such as trees and stones," he writes.

I think about the Tamil Sangam poets, who understood animal inner worlds and ecological grammar millennia before Western science.

# VIAVIAVIA

**I**T HAS BEEN MANY YEARS SINCE WE CONNECTED AND IN that time our waters have witnessed more violence and political turmoil," I write to my kalyanamitra. "Our wastewater treatment plants can now track a global virus. I have been thinking of you."

She replies that she, too, has been thinking about me. "Our life is very miserable, but I did not sit and suffer," she tells me. "I do water education for grassroots, ordinary women and conscious leadership for (potential) righteous leaders to be." She is looking for ways to continue to work "in the darkness."

I went to Burma searching for a lighthouse, for this love story between a person and an idea. I traveled the Irrawaddy from its source in Myitson to its mouth in Tavoy. What have I come to say?

In times of darkness and danger, we are always looking for light. The light comes from our convictions (like thek that lut), and our company (like kalyanamitra). I tell her she has always been that beacon of light.

When we speak next, she tells me this:

> You see, the coup is not government.
>     There is no fresh income for cash. So, what we do is cut the expenses.
>     We keep sharing our resources. . . . When time comes, we can earn some more money. . . . But when people are starving, and now the northern part, we have the cold wave, and people have no warm clothes, no blankets. It's about sharing.

She tells me about a catena of trust:

> Our currency is trust. So, if I trust you, I tell you everything. And then you tell your trusted friend, and then that trusted friend cannot see me, and they don't know me, but they have a trust in you. It's called ViaViaVia, you see. Our chain of trust makes things happen.

And as always she brings it back to water:

> I'm able to dream but unable to talk. What we really need is to keep our rivers and our water resources clean. We also need infrastructure.
>     This looks like a jigsaw to me. I have all the pieces, but they are just on the table. I have to place them to paint a picture. This is a struggle of a woman trying to get water and peace.

# SANGAM OF SANGAMS

**M**Y OTHER KALYANAMITRA, LISA, INVITES ME TO COLLABORATE on a project at the Toronto Waterfront, where the mouth of the Don River is being reconstructed. The river was rerouted and covered, she tells me. Water was filled with earth. The earth was poisoned. The river's mouth will now be restored. Contaminated soil will be removed to shape an island. Beneath layers of soil, Lisa tells me, seeds from pre-industrial times were found. We wonder if they can germinate and flourish now.

Lisa wants to create welcome signs for this new land near the new river mouth. They are *poetic infrastructures*, or as she calls them, "Careful Infrastructures for these reassembled lands."

I think about these dormant seeds and about water—both carry histories and shape futures. What does the river's mouth want to say?

*Men may come and men may go, but I go on forever.*

In these reassembled lands, new marsh plantings will emerge to sustain the emerald shiner, green frog, painted turtle, belted kingfisher, red fox.

What do we want to say to them? I am delighted to be invited to participate, to think about infrastructure not just through a technical or engineering lens, but a more philosophical one, to see infrastructure as the underlying framework for a society.

I am excited for this project to restore the mouth of a river. The mouth of a river is not where she begins, but where she ends. But a river does not end, she arrives at her confluence. She gives all she is and all she has carried to the sea, at this sangam of sangams.

These poetic infrastructures remind me of King Ashoka's rock edicts, and how his teachings on compassion and tolerance are etched in stone and sprinkled through the vast empire.

The edicts were also about the importance of public works. They prescribed that banyan trees should be planted alongside roads to provide shade to humans and other animals; medical stations should be placed for humans and other animals where medicinal gardens would also be planted to provide herbs with healing properties. Mango groves and water wells should be planned along roadways. These edicts were the ethical infrastructure of his society.

I return to how Lisa and I met. We were at a different waterfront, at an artist residency in the south of France, traipsing the limestone cliffs of a newly established national park. Peaceful coexistence with our wildlife neighbors has been something we both value, and we have continued to share stories from our respective cities. Over the years, we have had other conversations about how different cultures navigate relationships with wildlife. How they view them as relatives, and their presence is a blessing and not a nuisance.

I shared with her an article, "Indigenous insights on human–wildlife coexistence in southern India," published in *Conservation Biology*. From this, I learn the Kattunayakans, an Adivasi community in Kerala, refer to the wild animals with whom they share the forest as bandhukal (*relatives*), swantha ala (*our own people*), bhudiulla jeevikal (*intelligent beings*), and daiva (*God*). The forest is acchan-amma (*father mother*). The wetlands are petta amma (*birth mother*). *Who is responsible for the suffering of your mother?*

The researchers collect oral histories:

"For us, we want animals in the forest. We do not want them to be captured and taken away. If we see animals every day, there is another set of happiness."

"When we see one, we will bow and remember our gods in our hearts. Move away from their path, and we both go our ways."

They share with and give dharmam (*alms*) to their forest kin.

"Even if humans, dogs, or chicken, everyone should do their sacred duty—giving dharmam. Whatever we eat, it should be shared with others."

～

So much of our ecological crisis stems from a violent relationship with nature, and so much of that harmful relationship stems from a scarcity mentality, or fear of losing something. What if we turn to ViaViaVia and a currency of trust? I return to Williams's catena and invoke Seneca relaying Sotion relaying Pythagoras, who probably learned this from those Egyptian priests: "And what loss have you in losing your cruelty?"

Consider what we gain when we do. I think again of the astragale de Marseille, and her radical giving.

At first, I think I am making an homage to Ashoka's edicts, but perhaps I'm making an homage to his daughter, my namesake, with my own Sangamithra edict, for my own invisible city, my submerged Kumari Kandam, my contribution to the catena that came before me and the one that will come after—my Kallakurichi.

My Kallakurichi is a philosophy, a love story, a poem, a phoenix, a kolam, my government edict:

Swim. Fly. Nest.

Gather. Love. Breathe.

All Species Welcome.

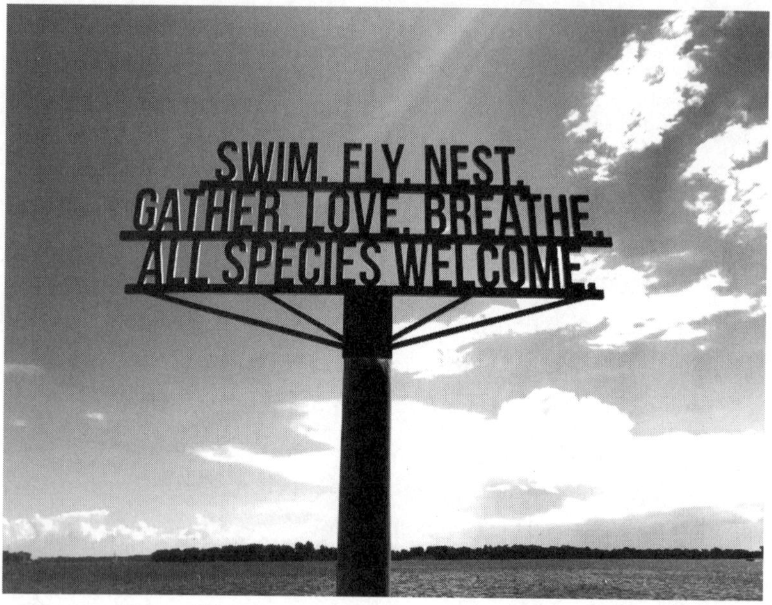
SWIM. FLY. NEST.
GATHER. LOVE. BREATHE.
ALL SPECIES WELCOME.

# ACKNOWLEDGMENTS

## MY KALYANAMITRA

I love acknowledgments because they make visible so many invisible labors and kindnesses that allow our work and our lives to flourish. I have been writing for so long that I've had many fellow travelers on my journey—so many kalyanamitra.

This is a book born in loss, and I want to begin by acknowledging the eternal. My father, Adi, whose death started this quest, and who kept me company the whole way—*this is our kind of book*.

To all the family I reconnected with after this loss, thank you for sharing your stories. Romba thanks to Jaya Aunty, G. S. Uncle, and Annu for guiding us to Ganga. I treasure Jaya Aunty's *jolly jolly* wisdom and G. S. Uncle's puns. Amirthalingam Uncle and Sarojini Aunty, thank you for the great gift of bringing me to Kallakurichi. I am grateful for your laughter and stories. To Savi Aunty and Dinesh Uncle, thank you for the family—all the tickets—you provided in this hemisphere. To Uma Aunty, thank you for your wellspring of joy, and to Krishnan Uncle, thank you for your remembrances. Sachi Aunty and Rajgopal Uncle, I am thankful for the stories you shared about my parents' meeting. To all the children of Kallakurichi, I will always be drawn to your lives and your stories. To my cousins of Kallakurichi, I am so happy we are connected.

Sending my deep love to my mother, Raji, for being an exemplary, kind, and generous soul, and mikka nandri for her support as my Tamil translator and keeper of memories. When I mapped out the table of contents as a watershed, she immediately recognized it: "That's great, Sangu! It is Triveni—the confluence of three rivers: Ganga, Yamuna, and the invisible Saraswati. I am happy for you." Thank you to my brother, Manu, another beautiful soul who aims to spread joy and kindness in a world that desperately needs it.

Although this book focuses on the paternal side of my family, I am deeply grateful for the wisdom and love on my maternal side. To Nana Thatha and Apeetha Patti, your home was my earliest travel destination. Thank you to Ranjini Aunty and Kala Aunty for encouraging my writing, and to Lalitha Mami, who helped with some early research. To the Manickams—Hema, Ashwath, Abhi, Lavi, the Linden extension: Arvin, Sareekha, and Peri, and the Canada contingent—thank you for being with me throughout this process. Lavanya Manickam, I am beyond thrilled to collaborate with you on the design. Thank you for making my sangam dreams come true.

Gamsahamnida to Eomeoni and Abeoji, and to my sisters-in-law and nieces for your encouragement.

My late mentor Louise DeSalvo once wrote, "The creative act has been misconstrued by many as a solitary and solipsistic act. [. . .] We must write about the creative act as it is nurtured by loving friends." I have been fortunate to be part of so many nourishing activist and writing communities.

To my *Satya* fam—Beth Gould, Catherine Clyne, Rachel Cernansky, Kymberlie Adams Matthews, Eric Weiss, and Mo Wyse—*Satya* was my first home for my writings, and you were my kin in caring. You helped me realize the writer I wanted to be: curious, expansive, and generous. Rachel, thank you for being my travel companion in Rwanda and in life. Beth, working on *The Long View* with you is one of my most cherished experiences. Cat, I have always admired your keen insight and righteous compassion. Kym, thank you for the deep dives into animal rescue and vegan cake. Eric and Mo, your passions have always energized me. I am also grateful to Christine Morrissey, one of our contributing authors who continues to inspire me with the myriad ways she cares for animals and her community.

Nancy Rawlinson may not remember, but I was in a Sackett Street workshop with her, trying to figure out my life after *Satya*. I shared a few pages of work, and she told me, "You know, there's enough here for a whole book." For planting that seed, thank you.

To my Hunter College MFA community: I am grateful for the time we spent holding one another's stories with grace and generosity. Many thanks to: Louise DeSalvo, who reminded us that "memoir is a corrective to history" and never to negotiate against ourselves; to Meena Alexander, who understood the necessity of fragments; to Kathryn Harrison, for your gentle guidance; and to Suketu Mehta, for your mentorship and encouragement.

To my Brooklyn Salon—Beth Gould, Adrienne Jones, Emily Bass, Lisa Freedman, and Jennifer Lutton—thank you for our many conversations and meals around the table, holding space for our collective radical works of care and justice. Arms up!

Thank you to Mia MacDonald of Brighter Green for the opportunities and support in researching the globalization of factory farming and for all the work you are doing to combat it. Thank you to Martin Rowe of Culture and Animals Foundation for your support of my thinking and curating of writings about animals.

To Rahna Reiko Rizzuto and Elena Georgiou, thank you for the generative radiance at Pele's Fire—I am still holding on to that glow. And to Bhanu Kapil, who guided us from afar to find the verge.

I am thankful for summer retreats at the Norman Mailer Writers Colony in Provincetown, led by William Lennon, Charles Hosier, and Amitava Kumar, and for my classmates and our time together by the water. Thank you, MaryBeth, for taking me to see Mary Oliver.

Thank you to Marie Chan, Adrienne Broder, and Aspen Summer Words for the Emerging Writers Fellowship. Aran Shetterly,

ACKNOWLEDGMENTS

your generosity and insight helped me see the ambition and heart of this book. To my Aspen classmates, thank you for sharing your amazing books with me.

My deep thanks to Joey McGarvey—a judge that year—who reached out with appreciation of the connections I was making with water, engineering, animals, and family history, and for later acquiring the book and believing in its expansiveness.

Thank you to Minal Hajratwala for connecting me to an essential writing community.

Geeta Kothari and Parul Kapur, it has been so wonderful to have you both as anchors in my writing life and to share updates with you week by week and year by year. Geeta, thank you for the myriad of ways you have championed this book. Thanks to my Unicorn comrades who provided community during the isolating days of the pandemic: JoAnn Balingit, Mahi Palanisami, Sunu Chandy, and Jen Soriano, it has been a privilege to be in process with you. M. Jacqui Alexander, thank you for our connections on the bounty and mystery of water. Leila Nadir, grateful to have you as my vegan, animal rescue, writing buddy. Leslie Tucker—my bosom friend—your abundant generosity and friendship have made this book all the richer. I'm in awe of the magic that happens when we discuss our books and our lives with each other.

To Edvige Giunta, who invited me into her hundred-words group with Tracy Mann, Celia Bland, Mary Giaimo, William Web, Dorothy Calbertini, and Julija Sukys, I am so happy to receive your words every day. To my other hundred-words group: Leslie Tucker, JoAnn Balingit, M. Jacqui Alexander, Jenne Patrick, Jen Soriano, and Maria Ressa, I am grateful to have built book fragments with you.

Thank you to Q. M. Zhang, for being an excellent reader and thought partner as I navigated the middle of this book. You reminded me that "we think of memory as a creature of the past, yet it lives in the present."

Thank you to Monique Holt, who taught me ASL, and to my professors at Cooper Union and UC Berkeley, whose soil and water lessons turned out to be life lessons. Thank you to Roger Fouts and Mary Lee Jensvold, who made a formative impression on me as a college student and set me on the path of chasing apes. To Sheri Speede and the Sanaga Yong Chimpanzee Rescue Center: Caring for baby chimpanzees in Cameroon has been one of my greatest honors. Thank you to Felix Akorli in Rwanda and Toby Cumberbach for making the introduction.

Thank you to Nitin Goel, Rakesh Bhardwaj, and Chetana Mirle for your guidance on animal industries in India. Naisargi Davé, thank you for your writing and thinking on these subjects, for reading early drafts of the chicken and cow chapters, and for pushing me further.

Thank you to David Naimon, whose podcast accompanied me on so many walks and train rides—and provided nourishment and inspiration for this literary life.

To Madhur Anand, thank you for inviting me to the Guelph Collaboratory; Forrest Gander, I cherish our conversations on the Sangam poets there, and thank you for introducing me to Chitra Sankaran's work; Brad Morrow, thank you for including me in *Conjunctions: Ways of Water*.

Urakose cyan to T, my Kinyarwanda translator: I treasure the work we did together and acknowledge you in my heart.

Mikka nandri to Sasipriya, my Tamil tutor and to Srinivasan Uncle for final Tamil fact checks.

I am grateful for brilliant friends who help me think about the big and small sorrows and joys of the world.

Hug/love to Valerie Chalcraft, my animal activist friend and mentor across so many decades, cities, and time zones.

To Lisa Hirmer: Thank you for our radiant friendship and our animal coexistence conversations and collaborations in Cassis, Queens, and Toronto. I look forward to many more.

Mikelle Adgate, my water buddy and visioning-the-world-we-want buddy, thank you for your generous readings and for your writings.

ACKNOWLEDGMENTS

To Iselin Gambert: Thank you for holding space for grief and joy with me and for dreaming of Land and Monkeys together.

Katie Hoek and Ruby Wells, grateful for our chats, walks, drives, and meals together.

To Shruti Swamy: What a gift your pen pal-ship has been. I am continually comforted and moved by your words.

To Sanjay Iyer: Thank you for reaching out and offering to be my dowsing guide.

To Mar: Thank you for dreaming of writing books together when we were young.

To John: Thank you for being my perfect reader and thinking buddy. I love cocurating ideas, meals, and gatherings with you.

For my Kalyanamitra NiNi, thank you for being my water guide to Burma.

To my Friends of Cooper Union crew—Rocco Cetera, Xenia Diente, Karina Tipton, Kerry Carnahan, and Henry Chapman: Our work correcting the historical record and imagining *The Way Forward* taught me so many lessons that I applied to researching this book, and I'm grateful to have learned them with you.

To Jen Kabat, thank you navigating the book process with me.

To my many passionate colleagues over the years doing the invisible work of water, animal, and environmental protection: I see you and thank you.

This book has been enriched by the support of a Jerome Foundation Study / Travel Literature Grant, which allowed me to visit the British Library in London and trace my grandfather's footsteps in Burma. Cindy Gehrig provided useful guidance during that time, which I still reflect on today. I am also grateful for a Jerome Foundation–supported residency at the Camargo Foundation in Cassis, which inspired the chapter "Sanctuary City." Merci to Julie Chenot at Camargo, Francis Talin, Lidwine Le Mire Pecheux, Patrick Bonhomme, Jeremy Boessau at Calanques National Park, Isabelle Laffont-Schwob and Alma Heckenroth at Aix-Marseille University, Baptiste Lanaspeze at Wild Project, and Gilles Clément for our encounters in France.

## ACKNOWLEDGMENTS

My deep gratitude to the Whiting Foundation and their Creative Nonfiction Grant, which afforded me the time to take this book to the next level. I am also grateful to the Café Royal Literature Grant for funding critical elements of travel, research, and translation.

My literary home away from home has been the Shoichi Noma Room of the New York Public Library. I am thankful to the Diamonstein-Spielvogel Fellowship, which granted me access to invaluable texts and the time to allow my sources to start "talking to each other." Special thanks to Jackie York and Rebecca Federman at the NYPL for accompanying me on my deep dive into PL-480 and helping me understand how my sources found their way to my shelves.

To the whole team at Milkweed Editions for supporting transformative literature: I am so grateful for you and your belief in this book. Thank you to Daniel Slager for completing this journey with me; to Tessa Carvalho and Adi Gandhi for your close eyes on this; to Lauren Langston Klein, for your gentle guidance through the publishing process; to Mary Austin Speaker and Mike Corrao for our design collaborations and your artistic vision; to Morgan LaRocca, Craig Popelars, Natalie Wollenzien, and Katie Hill, for your efforts in getting this book to readers out in the world.

To my agent, Anjali Singh: You've been my emotional support animal throughout this journey. Thank you for being my friend and advocate. I am grateful for our first meal of idlis together and all that has come since.

I started this book with Moo Cow by my side and finished it with Asta—two of my greatest loves. Thank you to our doggie caregivers—Karen, Tammy, Jon, Hema, Beth, Michael, Michelle, Veronica, and Germain—for watching over them while I traveled and worked on this book.

For the small smalls and all the animals whose stories populate this book: I continue to work for a world where your kin are safe and thriving.

Much of this book was written and revised traversing the NYC subway on Munsee Lenape lands.

## ACKNOWLEDGMENTS

There is the book, and there is the making of the book, which is also the making of a life.

To Wan: my champion, collaborator, fellow traveler, excellent reader, and co–pit bull parent extraordinaire. You continue to inspire and amaze me, and I love the life we've created together. Let's go places.

This book begins and ends with water. Following water has helped me understand what this book wants to be. Water is witness, memory, history, and life. It's what I've been divining all along.

---

Earlier versions of some portions of this book have been previously published in the following journals and anthologies under these titles:

*The Kenyon Review*: "Governing Bodies"
*Sister Species: Women, Animals, and Social Justice* (University of Illinois Press): "Small Small Redemption"
*Hippocampus Magazine*: "Pile Driving, Poetry and Poultry"
*Creative Nonfiction*: "Burmese Years"
*Conjunctions*: "Letter to My Submerged Father"
*Local Knowledge*: "Fingerprint of Water"
*Satya*: "The Long View"
*Primate People: Saving Nonhuman Primates Through Education, Advocacy, and Sanctuary* (University of Utah Press): "Soiled Hands"
Thrash Press: "Disenfranchised Grief in Four Parts," as a broadside
*Newtown Literary*: "Trying"
*1001 Nuits*: "Ingénieurs d'écosystèmes"

This book has also been informed by other writing I have done for *Writing for Animals* (Ashland Creek Press), *Letters to a New Vegan* (Lantern Publishing & Media), *Satya: The Long View*, and *VegNews*.

# THE WELL
## A CATENA OF SOURCES

### I. Divining Water

#### Divining Water

Earlier excerpts from this section were first published in *The Kenyon Review*: Sangamithra Iyer, "Governing Bodies," *The Kenyon Review* 41, no. 1 (January/February 2019).

5      ***these lines of Alfred, Lord Tennyson's***   Alfred, Lord Tennyson, "The Brook," *Memory Work and Appreciation*, ed. Ernest J. Kenny (Edward Arnold and Co., date unknown). I memorized by heart these lines published in a worn copy of *Memory Work and Appreciation*. Much of my book is also a practice of memory, work, and appreciation. A compilation of memories, tidbits learned over time, some of which predate the writing of this book, and others encountered during the process of it.

8      ***I write that story for my father***   Sangamithra Iyer, "A Sanctuary for an Orphaned Forest," *Satya*, September 2003.

10     **Harijan Welfare Worker**   The term *Harijan (person/ child of God)* was popularized by Gandhi to replace *untouchable*. Those working on their behalf and to eradicate untouchability were part of Harijan Sevak Sangh and considered Harijan Welfare Workers. The word *Dalit (broken, split)* was advocated for by B. R. Ambedkar. *Harijan* is used in this text to refer to Gandhi's newspaper, *Harijan*, and where it is used in original source material/context, and *Dalit* is used elsewhere. For more on Gandhi's choice of *Harijan*, see "Why 'Harijan'?" in the February 11, 1933, issue of the magazine, where Gandhi explains, "It is not a name of my coining. Some years ago, several 'untouchable' correspondents complained that I used the word 'asprisya' in the pages of [Gandhi's earlier journal] *Navajivan*. [. . .] One of the 'untouchable' correspondents suggested the adoption of the name 'Harijan' on the strength of its having been used by the first known poet-saint of Gujarat."

11     **etched into my memory**   William Lennon Jr., "To Become a Yogi" (1970). This is a self-published booklet dedicated to M. N. Adinarayanan.

## On Bridges

I owe my ability to pursue purpose-filled endeavors to a tuition-free education at The Cooper Union for the Advancement of Science and Art. I wrote more about its importance for *n+1*, available at https://www.nplusonemag.com/authors/iyersangamithra/. Thanks to Professor Guido for his introduction to geotechnical engineering.

19     **the local paper in Rockland County**   Peter W. Sluys, "Academic Revolution Coming to N. Rockland: Athletes Are Scholars; Scholar Athletes," *The Independent* 2, no. 28 (December 21, 1994).

20   *interest in the New York Liberty*   For more on the inaugural WNBA season of 1997, visit https://liberty.wnba.com/team-history/.

22   *primatologist Roger Fouts's memoir*   Roger Fouts and Stephen Tukel Mills, *Next of Kin: What Chimpanzees Have Taught Me About Who We Are* (William Morrow, 1997).

22   *DIRTY-GOOD is her phrase*   Fouts and Mills, *Next of Kin*, 82.

22   *swan as WATER BIRD*   Beatrix Gardner and Alan Gardner, *Teaching Sign Language to Chimpanzees* (State University of New York Press, 1989), 81.

24   *Washoe understands this too*   Fouts and Mills, 291.

## Soil Mechanics

In this chapter I also include teachings from my geotechnical engineering professors at UC Berkeley: Ray Seed, Juan Pestana, and Nick Sitar.

28   *a biography of Karl Terzaghi*   Richard E. Goodman, *Karl Terzaghi: The Engineer as Artist* (The American Society of Civil Engineers, 1999).

28   *When Terzaghi first arrives*   Goodman, *Karl Terzaghi*, 39–40.

28   *Terzaghi told his college students*   Goodman, *Karl Terzaghi*, 2.

28   *Pan for possibility*   Louise DeSalvo, personal communication.

28   *This book has no antecedent*   Goodman, *Karl Terzaghi*, 94.

28   *is credited to her invisible labor*   Goodman, *Karl Terzaghi*, 229.

28   *I am at a book reading*   "Migrant Father Fragment," Asian American Writers' Workshop conversation featuring Q. M. Zhang and lê thi diem thúy, August 17, 2017, https://aaww.org/aaww-tv-migrant-father-fragment/.

29 ***This ephemeral art*** Vijaya Nagarajan, *Feeding a Thousand Souls: Women, Ritual, and Ecology in India—An Exploration of the Kolam* (Oxford University Press, 2018), 4.

## In Situ

This chapter is drawn from my experiences as a geotechnical field engineer in California.

## Small Small Redemption

An earlier version of this chapter was originally published in *Sister Species: Women, Animals, and Social Justice*, edited by Lisa Kemmerer and published by the University of Illinois Press (2011).

More info on the Sanaga-Yong Chimpanzee Rescue Center is available at https://www.sychimprescue.org/.

Sheri Speede's memoir of Sanaga-Yong is also available for further reading: Sheri Speede, *Kindred Beings: What Seventy-Three Chimpanzees Taught Me About Life, Love, and Connection* (Harper Collins, 2013).

36 ***Vinoba talks about taking a vow of poverty*** Mahadev Desai, quoting Sjt. Vinoba in Weekly Letter, *Harijan*, March 14, 1936. *Harijan* spanned a critical time in Indian history leading up to and after WWII, starting from 1933 to 1955, with some periods of suspension during the war. The newspaper pursued independence not only from colonialism, but the tyrannies of industrialization, and asked what *nonviolence* meant in the age of the atomic bomb. Its audience was both Indian and foreign leadership, the rural masses; it was also for people like my grandparents, who left their lives in other cities to devote the rest of their time toward rural village rejuvenation, and it had readers and correspondents from all over the world.

38   ***Their body parts***   Dale Peterson, *Eating Apes* (University of California Press, 2003).

42   ***water-dowsing chimpanzees in Uganda***   Hella Péter, Klaus Zuberbühler, and Catherine Hobaiter, "Well-Digging in a Community of Forest-Living Wild East African Chimpanzees (*Pan troglodytes schweinfurthii*)," *Primates* 63 (2022): 355–64, https://doi.org/10.1007/s10329-022-00992-4.

43   ***I try to reassure her***   The Roots, "You Got Me," track 15 on *Things Fall Apart*, MCA, 1999.

43   ***do-it-yourself medical guide***   David Werner, *Where There Is No Doctor: A Village Health Care Handbook* (Hesperian Foundation, 1977). I mention the book *Where There Is No Doctor* on the shelf of the veterinary clinic in Cameroon. Years later I found this beautiful tribute to this book: Aminatta Forna, "Where There Is No Hospital," *Orion Magazine*, Spring 2022.

43   ***you also provide a similar role***   Mohandas Karamchand Gandhi, "Simple Cure for Scorpion Stings," *Harijan*, October 19, 1935.

44   ***the Chad–Cameroon pipeline***   More on its promises: The Associated Press, "Major Oil Pipeline Project Is Begun in Chad," *New York Times*, October 19, 2000. And its failures: Lydia Polgreen, "Chad's Oil Revenues Fail to Reach the Poor," *New York Times*, September 11, 2008.

44   ***paper on San Francisco Bay Mud***   Rudolph Bonaparte and James K. Mitchell, *The Properties of San Francisco Bay Mud at Hamilton Air Force Base, California* (Dept. of Civil Engineering, University of California, Berkeley, 1979).

## Pile Driving

An earlier version of this section was published in *Hippocampus Magazine*: Sangamithra Iyer, "Pile Driving, Poetry and Poultry," *Hippocampus Magazine*, January 1, 2013.

52   **the margins of my field log**   I sprinkle the poems and lines I scribbled on my field logs throughout this chapter.

## Burmese Years

Earlier excerpts from this section were first published in *The Kenyon Review*: Sangamithra Iyer, "Governing Bodies," *The Kenyon Review* 41, no. 1 (January/February 2019). Earlier excerpts about the process of uncovering this information were published in *Creative Nonfiction*: Sangamithra Iyer, "Burmese Years," *Creative Nonfiction* 56 (Summer 2015).

56   **an animated gorilla in a zoo**   *Creature Comforts*, dir. Nick Park (Aardman Animations, 1989).

56   **I pore through volumes of names**   I am grateful for the Jerome Foundation, who provided support for my research at the British Library in London, and to trace my grandfather's footsteps in Burma. I was able to track and trace my grandfather's civil service record in Burma by visiting the British Library and going through their Burma V/13 (yearly civil lists) series. I also referenced their annual files related to the public works department. Shout out to reference librarian John O'Brien.

57   **One compilation of elephant stories**   Lt. Col. J. H. Williams, *Elephant Bill* (Verdun Press, 2016).

60   **Emma Larkin, in her book**   Emma Larkin, *Finding George Orwell in Burma* (Penguin Press, 2005).

60   **As for the job I was doing**   George Orwell, "Shooting an Elephant," *New Writing* 2, Autumn 1936.

60   **In his novel Burmese Days**   George Orwell, *Burmese Days* (Oxford University Press, 1934).

61   **A kheddared elephant**   Williams, *Elephant Bill*, 56.

61   **In her biography of Williams**   Vicki Constantine Croke, *Elephant Company: The Inspiring Story of an Unlikely Hero*

and the Animals Who Helped Him Save Lives in World War II (Thorndike Press, 2015), 92.

62  **Williams claims his elephants**  Williams, *Elephant Bill*, 56.
62  **The oozie then sings to her**  Williams, *Elephant Bill*, 58–9.
63  **I pore through the transcripts**  I reviewed *Collected Works of Mahatma Gandhi* to locate Gandhi's speeches from his 1929 trip to Burma. His speech at a public meeting in Rangoon was reprinted in *Young India* on April 4, 1929.

## II. Our Kind of Book

Excerpted smaller fragments of this section were published in *Conjunctions* 80: *Ways of Water*: Sangamithra Iyer, "Letter to My Submerged Father," *Conjunctions* 80: *Ways of Water*, ed. Bradford Morrow, Spring 2023.

### Question Mark

I wanted to memorialize the names of the thirteen children of Kallakurichi, who continue to inspire more questions for me.

### Fingerprint of Water

An earlier abridged excerpt of this chapter was published in *Local Knowledge*: Sangamithra Iyer, "Fingerprint of Water," *Local Knowledge*, no. 1 (Spring/Summer 2013).

73  **The network of fluid-filled spaces**  Petros C. Benias et al., "Structure and Distribution of an Unrecognized Interstitium in Human Tissues," *Scientific Reports* 8, no. 4947 (2018): https://doi.org/10.1038/s41598-018-23062-6.
76  **I read about Shree Veer Bhadra Mishra**  Vandana Shiva, *Ganga: Common Heritage or Corporate Commodity*, (Navdanya, 2003), 8.
79  **trying to describe Varanasi**  Shiva, *Ganga*, 14.

## Air

82 **the breath of life** David Shulman, *Tamil: A Biography* (Belknap Press: An Imprint of Harvard University Press, 2016), 107.

82 **histories are often mixed with mythologies** Shulman, *Tamil*, 32.

## Without Sorrow

85 *The one about King Ashoka* Here I recount the story my father told me about Ashoka. I have also consulted several texts to learn more about Ashoka and my namesake Sanghamitra: George Kotturan, *Ahimsa: Gautama to Gandhi* (Sterling Publishers PVT, 1973); Nayanjot Lahiri, *Ashoka in Ancient India* (Harvard University Press, 2015); J. N. Samaddar, *The Glories of Magadha* (K. P. Jayaswal Research Institute, 1990).

87 *the vocabulary words of the chimpanzees* Beatrix Gardner and Allen Gardner, *Teaching Sign Language to Chimpanzees* (State University of New York Press, 1989).

87 *The chimpanzee Tatu's column of verbs* Gardner and Gardner, *Teaching Sign Language to Chimpanzees*, 208.

87 *Moja has two listed* Gardner and Gardner, *Teaching Sign Language to Chimpanzees*, 206.

87 *Sprinkled across Ashoka's vast empire* I consult several sources on Ashoka's Rock Edicts: Lahiri, *Ashoka in Ancient India*; Samaddar, *The Glories of Magadha*; E. Hultzsch, *Corpus Inscriptionum Indicarum Vol. 1: Inscriptions of Asoka, New Edition* (Indological Book House, 1969).

88 *an elephant named Tara* Mark Shand, *Travels on My Elephant* (Overlook Press, 1992), 41.

88   *akam and puram*   David Shulman, *Tamil: A Biography* (Belknap Press: An Imprint of Harvard University Press, 2016), 45.

88   *a conversation with the poet Arthur Sze*   David Naimon, host, *Between the Covers*, "Arthur Sze: The Glass Constellation: New & Collected Poems," Tin House, July 1, 2021, 2 hrs., 19 min., https://tinhouse.com/podcast/arthur-sze-the-glass-constellation-new-collected-poems/.

88   *as one Salman Rushdie character says*   Salman Rushdie, *The Ground Beneath Her Feet* (Henry Holt and Company, 1999), 19.

89   *chilling picture of a mountain of bison skulls*   Men standing with pile of buffalo skulls, Michigan Carbon Works, Rougeville Mich., 1892, photograph, Burton Historical Collection, Detroit Public Library, https://theconversation.com/historical-photo-of-mountain-of-bison-skulls-documents-animals-on-the-brink-of-extinction-148780.

90   *I wish I knew these lines*   Warsan Shire, "The Birth Name," *Fuck Yeah, Poetry!*, October 25, 2011, https://fypoetry.tumblr.com/post/34281180644/the-birth-name; Assétou Xango, "Give Your Daughters Difficult Names," Café Cultura, May 12, 2017, https://poets.org/anthology/poems-134.

91   **the city without sorrow**   Gail Omvedt, *Seeking Begumpura: The Social Vision of Anticaste Intellectuals* (Navayana Pub, 2008), 1.

91   **Utopias, the posing of alternatives**   Omvedt, *Seeking Begumpura*, 14.

92   **I think about this sorrow**   For more on han, I wrote this newsletter inspired by the scholarship of E. J. Koh: https://literaryanimal.substack.com/p/the-language-of-sorrow.

## Truth

97     *She writes about her first exposure*   Edward Maitland, *Anna Kingsford: Her Life, Letters, Diary, and Work* (Cambridge University Press, 2011). I quote this passage from the anti-vivisectionist Anna Kingsford, published in *The Heretic*. There is a brilliant novel, *Experimental Animals*, by Thalia Field that is written as an imaginary letter to Anna Kingsford.

99     *It's a terrific magazine*   Nellie McKay in reference to *Satya*'s Hurricane Katrina Issue, November 2005.

100     *Thirteen years later, Beth reflected*   Beth Gould, "Letter from the Publisher," *Satya*, June/July 2007.

100     *Martin declares in that introductory issue*   Martin Rowe, "Letter from the Editor," *Satya*, June 1994

101     *What might happen is*   Martin Rowe, "Editorial: Why You Oughta Know," *Satya*, August 1996.

101     *the psychotherapist Francis Weller*   Francis Weller, *The Wild Edge of Sorrow: Rituals of Renewal and the Sacred Work of Grief* (North Atlantic Books, 2015), 9.

102     *Cat details the anatomy*   Catherine Clyne, "Editorial: Anatomy of an Awakening," *Satya*, February 2000.

## The Chicken Issue

103     *our issue about Hurricane Katrina*   *Satya*, November 2005.

103     *Miyun Park, writes*   Miyun Park, "The Lucky Few," *Satya*, November 2005.

103     *The artist Sue Coe*   Leana Stormont, "Help Was Never on the Way," *Satya*, November 2005.

104     *Kym and Kristi take*   Kymberlie Adams Matthews, "Brooklyn Birds," *Satya*, February 2006. Kym has written a wonderful article detailing this experience of rescuing and finding homes for these birds.

107     *inspires an entire magazine issue*   *Satya*, February 2006.

## A Thing After Your Own Heart

109 *a promise to his mother* Mohandas Karamchand Gandhi, *Autobiography: The Story of My Experiments with Truth* (Navajivan, 1925–9), 39.

109 *As one friend explains* Gandhi, *Autobiography*, 20.

109 *doggerel of the Gujarati poet Narmad* Gandhi, *Autobiography*, 21.

110 *It began to grow on me* Gandhi, *Autobiography*, 21.

110 *His first experiment with this truth* Gandhi, *Autobiography*, 22.

110 *It is all very well* Gandhi, *Autobiography*, 42.

110 *But I have heard* Gandhi, *Autobiography*, 43.

111 *Rest assured it is a fib* Gandhi, *Autobiography*, 43.

111 *the Central Vegetarian Restaurant* Gandhi, *Autobiography*, 48.

111 *Salt begins with a confession* Henry S. Salt, *A Plea for Vegetarianism, and Other Essays* (Vegetarian Society, 1886), 7.

111 *while others admit the possibility* Salt, *A Plea for Vegetarianism*, 8.

112 *In truth, it is* Salt, *A Plea for Vegetarianism*, 19.

112 *articles Gandhi wrote in* **The Vegetarian** Mohandas Karamchand Gandhi, *The Collected Works of Mahatma Gandhi: Volume I*, (Navajivan Trust, 1958), 19–39. I consulted a number of Gandhi's early articles published in *The Vegetarian* in London, which are preserved in the Gandhi Heritage Portal. He wrote a series on Indian vegetarians, "Indian Festivals and Foods of India," published in 1891.

112 *like readers finding* **Satya** *in NYC* I wrote an editorial, "The Long View," for *Satya*'s 20th Anniversary Edition, where some of this recounting was previously published.

113 *satyagraha is coined in a contest* Joseph Lelyveld, *Great Soul: Mahatma Gandhi and His Struggle with India* (Alfred A. Knopf, 2011), 18.

113   **Gandhi befriends Henry Polak**   Lelyveld, *Great Soul*, 85.

115   **collection of Gandhi's letters**   B. Srinivasa Murthy, ed., *Mahatma Gandhi and Leo Tolstoy Letters* (Long Beach Publications, 1987). I read this collection of letters between Gandhi and Tolstoy, which Tolstoy received in the last year of his life, and Tolstoy's letters to Gandhi, which are among his last writings.

116   **call his first settlement Phoenix**   Mark Thomson, *Gandhi and His Ashrams* (Popular Prakashan, 1993), 48.

117   **Indian Opinion *encouraged readers to memorize***   Isabel Hofmeyr, *Gandhi's Printing Press* (Harvard University Press, 2013), 129.

117   **Thoreaus in miniature**   Hofmeyr, *Gandhi's Printing Press*, 131.

117   **this passage had a profound impact**   Hofmeyr, *Gandhi's Printing Press*, 90.

118   **an antidote to this bombardment of information**   Hofmeyr, *Gandhi's Printing Press*, 4.

118   **The influences are cross-cultural**   Wai Chee Dimock, "Global Civil Society: Thoreau on Three Continents," *Through Other Continents: American Literature Across Deep Time* (Princeton University Press, 2008).

119   **a change in United States immigration policy**   Rajgopal Uncle shared a personal story of how the news of the Immigration Act of 1965 traveled to India and resulted in my father's departure. He also shared with me the story of how my parents met.

121   **Mom interviewing you**   My mother recounted these questions to me.

122   **You write about the early freedom fighters**   M. N. Adinarayanan Iyer, "Punjab-Kesari Lala Lajpat Rai." This is an essay that my father wrote for a college contest and won.

122   **gifts this book to you**   Henry Casper, *Verdome* (1972).

124   **Casper had a wit and wisdom**   Casper, *Verdome*, 49.

## Love Is Work Made Visible

128 **moral injury** Rogé Karma, guest host, *The Ezra Klein Show*, "The Case Against Loving Your Job," *New York Times*, November 19, 2021, 1 hr., 23 min., https://www.nytimes.com/2021/11/19/podcasts/transcript-ezra-klein-show-sarah-jaffe.html. I had listened to this episode of *The Ezra Klein Show* which discussed moral injury and work. This episode was guest hosted by Rogé Karma with the guest Sarah Jaffe.

130 **When the chimpanzees make errors** Beatrix Gardner and Allen Gardner, *Teaching Sign Language to Chimpanzees* (State University of New York Press, 1989), 182.

## An Attitude of Gratitude

133 *no word for* **privacy** *in Hindi* Katherine Russell Rich, *Dreaming in Hindi: Life in Translation* (Portobello Books, 2011).

134 *Here are her sweet offerings* In a booklet titled *Tulasi Mata*, C. S. Rao compiled tributes to Jaya Aunty in both English and Tamil by her friends, family, and admirers. My mother assisted with Tamil translation.

## Work

139 *chimpanzee behavioral taxonomy* The Chimpanzee and Human Communication Institute provided a blue booklet for teaching us chimp behaviors.

## Think-Trouble

142 *who film and transcribe it* For more on Michael at The Gorilla Foundation, visit https://www.koko.org/about/gorilla-michael/.

## Soiled Hands

An earlier version of this chapter was originally published in *Primate People: Saving Nonhuman Primates Through Education, Advocacy, and Sanctuary*, edited by Lisa Kemmerer and published by the University of Utah Press (2013).

144     **one of the collective members suggests**   Bill Weber and Amy Vedder, *In the Kingdom of Gorillas: The Quest to Save Rwanda's Mountain Gorillas* (Simon & Schuster, 2002).

145     **I interview Carole Noon**   Carole Noon, "Opening the Door: Save the Chimps Gives New Life to Florida Retirement: The *Satya* Interview with Carole Noon," interview by Sangamithra Iyer, *Satya*, November 2000.

145     **in a sanctuary in Quebec**   Referring to The Fauna Foundation, https://faunafoundation.org/.

145     **On my daily commute**   I was reading *We Wish to Inform You That Tomorrow We Will Be Killed with Our Families* by Philip Gourevitch (Picador, 1999) on my Metro North rides to work.

150     **The gacaca in Butare**   I am grateful to the young man who sat next to me taking notes at the gacaca, and for my anonymous translator T, who worked with me page by page to discuss what is and isn't in these notes.

## Finding Kallakurichi

158     **unspoken suffering of Kasturba**   Mohandas Karamchand Gandhi, *Autobiography: The Story of My Experiments with Truth* (Navajivan, 1925–9), 31.

158     **being overrun with visitors**   Mohandas Karamchand Gandhi, "Apology to Visitors," *Harijan*, November 2, 1935.

160     **you "make clouds" by picking cotton**   Mahadev Desai, "Village Children's Poetry," *Harijan*, December 28, 1935.

"Making clouds" refers to how village children referred to making cotton. As Desai notes, "'Making clouds' is becoming quite a local joke and, you will agree, a pretty one."

160 **Spinning is Gandhi's answer**  Mohandas Karamchand Gandhi, "Therapeutic Value of Spinning," *Khadi (Hand-Spun Cloth): Why and How*, ed. Bharatan Kumarappa (Navajivan Publishing House, 1955).

162 **I played Mark Antony**  William Shakespeare, *Julius Caesar*.

163 **His debates with Ambedkar**  For more on Gandhi and Ambedkar, see Arundhati Roy, *The Doctor and the Saint: Caste, Race, and Annihilation of Caste* (Haymarket Books, 2017).

163 **will be outcastes**  B. R. Ambedkar, "Dr. Ambedkar & Caste," *Harijan*, February 11, 1933. In the very first issue of *Harijan*, Ambedkar provided this statement.

163 **Ambedkar challenges him**  Ambedkar, quoted in *Harijan*, May 9, 1936.

163 **he tests his Brahmacharya**  Joseph Lelyveld, *Great Soul: Mahatma Gandhi and His Struggle with India* (Alfred A. Knopf, 2011), 304.

163 **satyagraha relies on vulnerability**  Saurabh Dube, Sanjay Seth, and Ajay Skaria, eds., *Dipesh Chakrabarty and the Global South: Subaltern Studies, Postcolonial Perspectives, and the Anthropocene* (Routledge, 2020), 163.

164 **Sayana Uncle writes a memoir**  My mother had a Xerox of quotations that Sayana Uncle wrote by hand when he was stationed in Zambia.

164 **a poem your friend Del Rhodes**  Del Rhodes, "Always a Smile," given as a gift to my mother.

165 **Sayana Uncle writes this quote down**  Quote is attributed to Austin O'Malley.

165 **Ken Saro-Wiwa's last words**  Kenule Beeson Saro-Wiwa, "Ken Saro-Wiwa's Last Words," *Satya* Guest Editorial, January 1996.

165     *interview with Ken Saro-Wiwa's son*   Patricia Cohen, "A Writer's Violent End, and His Activist Legacy," *New York Times*, May 4, 2009.

171     **Aunty recites a Percy**   Percy Bysshe Shelley, "To a Skylark" (Charles and James Ollier, 1820).

172     **Gandhi recites a Shamal Bhatt poem to her**   Written by Shamal Bhatt, and recounted in Gandhi's autobiography: Gandhi, *Autobiography*, 35.

173     **For men may come and men may go**   Alfred, Lord Tennyson, "The Brook" (1886).

## Returns

175     *cities now submerged by the sea*   Sumathi Ramaswamy, *The Lost Land of Lemuria: Fabulous Geographies, Catastrophic Histories* (University of California Press, 2004), 117.

175     **become a mystery**   Charles Allen, *Ashoka: The Search for India's Lost Emperor* (Abacus, 2013).

175     **eradicated local knowledge of spinning**   One chilling account of British violence with respect to cotton can be found here: Sofi Thanhauser, *Worn: A People's History of Clothing* (Knopf Doubleday Publishing Group, 2022), 60–5.

175     **In Vijapur, she finds**   Mohandas Karamchand Gandhi, "Therapeutic Value of Spinning," in *Khadi (Hand-Spun Cloth): Why and How*, ed. Bharatan Kumarappa (Navajivan Publishing House, 1955). Account of learning to spin, first published in *Young India*, October 13, 1927.

175     **Charkha Spinning class in Brooklyn**   At Loop of the Loom in Dumbo.

176     **my Tamil tutor, Sasipriya**   I took lessons with Sasipriya over Zoom.

177     **an imagined conversation**   Italo Calvino, *Invisible Cities* (Harcourt Brace Jovanovich, 1978).

| | |
|---|---|
| 177 | **seizure by the sea**  Or "seizure by the ocean." Ramaswamy, *The Lost Land of Lemuria*, 117. |
| 178 | **the book T recommended**  Ka Pa Aṟavāṇan, *Anthropological Studies on the Dravido-Africans* (Tamil Koottam, 1980). |
| 178 | **A British zoologist called it Lemuria**  William Scott-Elliot, *The Lost Lemuria* (Theosophical Publishing Society, London, 1904). Philip Sclater coins the term, Lemuria. |
| 178 | **It is a lost place**  Ramaswamy, *The Lost Land of Lemuria*, 1. |

## Catena

| | |
|---|---|
| 179 | **indelible stories of Pythagoras**  Colin Spencer, "Pythagoras and His Inheritance," *The Heretic's Feast: A History of Vegetarianism* (University Press of New England, 1996), 33–68. |
| 180 | **I unbox a fragile volume**  Howard Williams, *The Ethics of Diet: A Catena of Authorities Deprecatory of the Practice of Flesh-Eating* (F. Pittman, 1883). |
| 180 | **his essay introducing the catena**  Leo Tolstoy, "The First Step," in *The Ethics of Diet: An Anthology of Vegetarian Thought*, ed. Howard Williams (White Crow Books, 2010), 45. |

## Verge

| | |
|---|---|
| 181 | **gives us instructions**  Bhanu Kapil's prompt at Pele's Fire retreat in Hawaii in 2017. |
| 181 | **interviews women of South Asian descent**  Bhanu Kapil, *The Vertical Interrogation of Strangers* (Kelsey Street Press, 2001). |
| 184 | **James Baldwin speaks about poetry**  James Baldwin, "The Black Scholar Interviews: James Baldwin," in *The Black Scholar* 5, no. 4 (1973–4): 33–42. |

185   ***focuses on the word depends***   William Carlos Williams, "The Red Wheelbarrow," in *The Collected Poems of William Carlos Williams, Volume I, 1909–1939* (New Directions Publishing Corporation, 1938). Thank you to Mr. Spaight, my AP English teacher, for instilling the question: "How does a poem mean?"

185   ***William Carlos Williams also writes***   William Carlos Williams, "Asphodel, That Greeny Flower," in *Asphodel, That Greeny Flower & Other Love Poems* (New Directions Publishing Corporation, 1955).

185   ***penned a passionate plea***   Rabindranath Tagore, "Self-Imposed Impoverishment," *Harijan*, January 11, 1936.

185   ***One poser writes***   Mohandas Karamchand Gandhi, "A Fatal Fallacy," *Harijan*, January 11, 1936.

186   ***fractal responsibility***   David Naimon, host, *Between the Covers*, "Crafting with Ursula: adrienne maree brown on Social Justice and Science Fiction," Tin House, May 10, 2022, 2 hrs., 39 min., https://tinhouse.com/podcast/crafting-with-ursula-adrienne-maree-brown-on-social-justice-science-fiction/.

186   ***what Jane Addams wrote***   Jane Addams, Introduction to *What Then Must We Do?* by Leo Tolstoy, trans. Aylmer Maude (Oxford University Press, 1942), xiii.

186   ***finds agency and power in daily life***   Mohandas Karamchand Gandhi, "A Fatal Fallacy," *Harijan*, January 11, 1936.

187   ***how he described partition***   Joseph Lelyveld, *Great Soul: Mahatma Gandhi and His Struggle with India* (Alfred A. Knopf, 2011), 298.

187   ***Ahimsa, too, was applied***   Mohandas Karamchand Gandhi, "The Greatest Force," *Harijan*, October 12, 1935.

188   ***truth was the first step***   Leo Tolstoy, "The First Step," in *The Ethics of Diet: An Anthology of Vegetarian Thought*, ed. Howard Williams (White Crow Books, 2010), 157.

| | |
|---|---|
| 188 | ***The Syrian poet Kahlil Gibran*** Kahlil Gibran, "The Prophet," *Harijan*, November 23, 1935. |
| 188 | ***Gandhi tells his new recruits*** Mahadev Desai, "A Talk to Village Workers," *Harijan*, November 2, 1935. Mahadev Desai provides a "gist" of Gandhi's talk to village workers. |
| 189 | ***the challenges of village rejuvenation*** Mohandas Karamchand Gandhi, "Snake Poisoning," *Harijan*, August 17, 1935. |
| 189 | ***in his quenchless thirst*** Mahadev Desai, "Snake Lore," *Harijan*, August 17, 1935. |
| 189 | ***easily accessible remedies*** Gandhi, "Snake Poisoning." |
| 189 | ***he decides to write Gandhi*** Mohandas Karamchand Gandhi introduces M. S. Narayanan's (Thatha's) "Simple Cure for Scorpion Stings," *Harijan*, October 19, 1935. |
| 192 | ***bees at Gandhi's ashram*** C. Rajagopalachari, "The Story of Our Bees," *Harijan*, January 25, 1936. |
| 193 | ***A reader named Rajaji writes in*** Rajaji, "Vermin-Proof Paper?," *Harijan*, December 14, 1935. |
| 193 | ***I write about divining*** Sangamithra Iyer, "Governing Bodies," *The Kenyon Review* 41, no. 1 (January/February 2019). |
| 195 | ***if the inferno*** Italo Calvino, *Invisible Cities* (Harcourt Brace Jovanovich, 1978), 165. |
| 195 | ***translations of these ancient poems*** A. K. Ramanujan, ed., *Poems of Love and War: From the Eight Anthologies and the Ten Long Poems of Classical Tamil* (Columbia University Press, 2011), x. |
| 196 | ***Swaminatha Aiyar, who encounters*** Ramanujan, ed., *Poems of Love and War*, xvi. |
| 196 | ***this Ramanujan passage*** Vijaya Nagarajan, *Feeding a Thousand Souls: Women, Ritual, and Ecology in India—An Exploration of the Kolam* (Oxford University Press, 2018), 1. |

| | | |
|---|---|---|
| 196 | ***as red earth and pouring rain*** | Ramanujan, ed., *Poems of Love and War*, ix. |
| 196 | ***contribute to an anthology*** | Melissa Tedrowe and Justin Van Kleek, eds., *Letters to a New Vegan* (Lantern Publishing & Media, 2015). |
| 197 | ***go into yourself*** | Rainer Maria Rilke, *Letters to a Young Poet*, trans. M. D. Herter Norton (W. W. Norton & Company, 1993). |
| 197 | ***my ulcerative colitis is*** | Antonina Mikocka-Walus, *IBD and the Gut-Brain Connection* (Hammersmith Books Limited, 2018). |
| 199 | ***The male humpback whales*** | At the time I was listening to a podcast, *Dolphin & Whale Tales, Wisdom from the Deep*, hosted by Anne Gordon. |
| 200 | ***the etymology of Tamil*** | David Shulman, *Tamil: A Biography* (Belknap Press: An Imprint of Harvard University Press, 2016), 4. |
| 200 | ***created a system of eris*** | Erica Gies, *Water Always Wins: Thriving in an Age of Drought and Deluge* (University of Chicago Press, 2023), 143. |
| 200 | ***We don't know much about Tiruvalluvar*** | Shulman, *Tamil*, 95. |
| 201 | ***I collect several versions*** | I have consulted several copies of *Tirukkural* in translation: Thomas Hitoshi Pruiksma, trans., *The Kural: Tiruvalluvar's Tirukkural* (Beacon Press, 2022); P. S. Sundaram, trans., *The Kural* (Penguin Books, 1991); Meena Kandasamy, *The Book of Desire* (Galley Beggar Press, 2023). |
| 201 | ***like a principle of* Satya** | Sundaram, trans., *The Kural*, 44. |
| 201 | ***in praise of rain*** | Pruiksma, trans., *The Kural*, 4. |
| 201 | ***These remind me of you*** | Sundaram, trans., *The Kural*, 78. |
| 201 | ***Sangam poets wrote about nature*** | M. Varadarajan, *Treatment of Nature in Sangam Literature* (The South India Saiva Siddhanta Works Publishing Society, 1957). |

| | | |
|---|---|---|
| 201 | *trees are intertwined with the species* | Varadarajan, *Treatment of Nature*, 259–60. |
| 202 | *the vênkai tree* | Varadarajan, *Treatment of Nature*, 266. |
| 202 | *monkey has clandestine union* | Varadarajan, *Treatment of Nature*, 398. |
| 202 | *the pleasures of a monkey* | Varadarajan, *Treatment of Nature*, 284. |
| 202 | *when a flock of cranes* | Varadarajan, *Treatment of Nature*, 285. |
| 202 | *roots of the irri tree* | Varadarajan, *Treatment of Nature*, 285. |
| 203 | *as feeling afraid* | Varadarajan, *Treatment of Nature*, 279. |
| 203 | *when the male elephant slips* | Varadarajan, *Treatment of Nature*, 281. |
| 204 | *mistake for a snare* | Varadarajan, *Treatment of Nature*, 281. |
| 204 | *teach us about animal grief* | Varadarajan, *Treatment of Nature*, 287–8. |
| 204 | *cries aloud piteously* | Varadarajan, *Treatment of Nature*, 307. |
| 204 | **Intergovernmental Panel on Climate Change** | IPCC, *Climate Change 2022: Impacts, Adaptation, and Vulnerability* (Cambridge University Press, 2023), https://doi.org/10.1017/9781009325844. |
| 204 | **Sangam poets communicated this precarity** | Varadarajan, *Treatment of Nature*, 286–7. |
| 205 | *his indelible description* | Varadarajan, *Treatment of Nature*, 286. |
| 205 | **2009 riverbed discovery** | A geologist team led by Muthuvairavasamy Ramkumar discovered dinosaur eggs in Tamil Nadu. *Nature India*, "Dino Eggs Treasure Trove Unearthed," September 30, 2009, https://doi.org/10.1038/nindia.2009.306. |

## The Truth Is in the Body

207  **paid equal attention to shade**  Xavier S. Thani Nayagam, *Nature in Ancient Tamil Poetry, Concept and Interpretation* (Tamil Literature Society, 1953), 12.
207  **Rain clouds accumulate the wealth**  Nayagam, *Nature in Ancient Tamil*, 16.
207  **A tender heart**  Nayagam, *Nature in Ancient Tamil*, 17.
208  **Will it be katai**  Sumathi Ramaswamy, *The Lost Land of Lemuria: Fabulous Geographies, Catastrophic Histories* (University of California Press, 2004), 154. The distinction between story, myth, and history applies to Lemuria and Kallakurichi.

## III. Ungovernable Bodies

### Triveni Sangam

213  **I return to Howard Williams's**  Howard Williams, *The Ethics of Diet: An Anthology of Vegetarian Thought* (White Crow Books, 2010), 83.

### Dead Zone

215  **Satya *published this spread***  "Actual Size," *Satya*, February 2006.
216  **Seneca lamented this globalization**  Howard Williams, *The Ethics of Diet: An Anthology of Vegetarian Thought* (White Crow Books, 2010), 84–5.
216  **I would reread the letters**  We published letters from our readers in last issue of *Satya*, June 2007. Letters are available at http://www.satyamag.com/jun07/letters.html.
217  **I interviewed Chetana**  Chetana Mirle was working for Humane Society International at the time. Grateful for

our initial conversations, and her invitation to meet the poultry barons in June 2008.

217 **The only poultry barons** John Steinbeck, *The Short Reign of Pippin IV* (Penguin Publishing Group, 2007).

## Another Dead Zone

I want to acknowledge support from Brighter Green for the documentary work detailed in this chapter and the following ones set in India. This research informed some case studies and policy papers published by Brighter Green. For more info, visit https://brightergreen.org/india/. I have also written about some of these topics in *VegNews* (November/December 2017).

Thank you also to Nitin Goel and Rakesh Bhardwaj for being the guide to these animal industries.

223 *small organic bokbunja* Bokbunja is technically wild black raspberry, but commonly referred to in the family as blackberry.

228 *do we need to see violence* I write about the challenges of bearing witness. I want to acknowledge Kathryn Gillespie's detailing of her own struggles with bearing witness in *The Cow with Ear Tag #1389* (University of Chicago Press, 2018). She promotes "embodied practices that honor the deeply emotional and often traumatizing nature of research dedicated to uncovering injustice and structural violence." Gillespie employed what she calls "buddy system research." She notes she and her buddy serve as companions who "accompany each other into field and analysis research processes that are emotionally taxing, potentially physically unsafe, or otherwise troubling to encounter alone." I am grateful to my husband Wan, for being my buddy into these dead zones.

## Bas Bas Bas

233     ***radical retelling of the epic***    Karthika Naïr, *Until the Lions: Echoes from the Mahabharata* (Steerforth Press, 2019).

237     ***Steinbeck's Egg King of Petaluma***    John Steinbeck, *The Short Reign of Pippin IV* (Penguin Publishing Group, 2007), 66.

## In Search of Sacred Cows

247     ***Calvino story about stubborn cats***    Italo Calvino, *Marcovaldo, or The Seasons in the City* (Harcourt Brace Jovanovich, 1983).

250     ***visited a flaying yard in Bombay***    Satish Chandra Das Gupta, "Carcass Disposal and Leather Trade," *Harijan*, November 30, 1934.

## The First Concept

253     ***Mother, my first sound***    Theresa Hak Kyung Cha, *Dictee* (University of California Press, 2009), 50.

253     ***Mother tongue is your refuge***    Cha, *Dictee*, 45.

254     ***The rupture of elephant lands***    G. A. Bradshaw, *Elephants on the Edge: What Animals Teach Us About Humanity* (Yale University Press, 2009), 43.

254     ***A matriarch's death***    Bradshaw, *Elephants on the Edge*, 11.

## Ruminate

255     ***Jain concept of saptabhanginaya***    A variation of the sevenfolds of truth can be found here: Nagin J. Shah, ed., *Jaina Theory of Multiple Facets of Reality and Truth (Anekantavada)* (Motilal Banarsidass, 2000).

256  **what Gandhi wrote**  Mohandas Karamchand Gandhi, *The Collected Works of Mahatma Gandhi: Volume I*, (Navajivan Trust, 1958), 19–39.

256  **Consider this account**  Yamini Narayanan, *Mother Cow, Mother India: A Multispecies Politics of Dairy in India* (Stanford University Press, 2023), 111–2.

257  **As B. R. Ambedkar noted**  Vasant Moon, ed., *Dr. Babasaheb Ambedkar: Writings and Speeches* (Education Department, Govt. of Maharashtra, 1979), 9, 195.

258  **refusing to dispose of cow carcasses**  Gopi Maniar Ghanghar, "Gujarat Dalits continue to protest, refuse to pick cattle carcasses from roads," *India Today*, July 28, 2016.

258  **a tanner set free**  Gail Omvedt, *Seeking Begumpura: The Social Vision of Anticaste Intellectuals* (Navayana, 2008).

258  **this line from a character**  Karen Joy Fowler, *We Are All Completely Beside Ourselves* (Putnam's Sons, 2013), 232.

258  **the couplets of** Tirukkural  P. S. Sundaram, trans., *The Kural* (Penguin Books, 1991), 78.

For more information on human and animal entanglements in India, I highly recommend Naisargi Davé, *Indifference: On the Praxis of Interspecies Being* (Duke University Press, 2023); and Yamini Narayanan, *Mother Cow, Mother India: A Multispecies Politics of Dairy in India* (Stanford University Press, 2023).

## Landscapes of Grief

259  **scholar Chitra Sankaran tells us**  Chitra Sankaran, "Retrieving the Margins: Use of *Thinai* by Three Contemporary Tamil Women Writers," *ISLE: Interdisciplinary Studies in Literature and Environment* 29, no. 2 (2022): 443–65, https://doi.org/10.1093/isle/isaa141.

259  **Each physical landscape**  Sankaran, "Retrieving the Margins," 443–65.

## My Kalyanamitra

263     **Gandhi asks the audience**    Mohandas Karamchand Gandhi, "Speech at Paungde," March 15, 1929.

263     **letter of award for a literature grant**    The research trips to the British Library and to Burma in 2013 were supported by a Travel/Literature Grant from the Jerome Foundation.

264     **riveted by a revenue report**    Proceedings of the Chief Commissioner, British Burma, in the Revenue Department, "Resolution on the Report on Light-Houses off the Coast of British Burma for 1877–78," India Office Records, British Library, July 12, 1878.

265     **pointed me to an index**    Proceedings of the Government of Burma, Public Works Department, 1920–4.

265     **What I can find about Rangoon's**    In the National Archives in Yangon, I was flipping through an 1878 report on the first "Rangoon Lunatic Asylum."

269     **Later, in our hotel, I read**    The Economist, "A Burmese Spring," *The Economist: Special Report*, May 23, 2013.

269     **In 1998 he came to Rwanda**    I first heard about Bill Clinton's non-apology in Rwanda here: Samantha Power, *"A Problem from Hell": America and the Age of Genocide* (Basic Books, 2013).

270     **one aqueduct that is leaking**    "Delaware Aqueduct Rondout – West Branch Tunnel Repair," NYC Environmental Protection, accessed April 23, 2025, https://www.nyc.gov/site/dep/about/delaware-aqueduct-rondout-west-branch-tunnel-repair.page.

272     **Time *magazine cover story***    Hannah Beech, "The Face of Buddhist Terror," *Time*, July 1, 2013.

275     **a lighthouse aficionado**    Russ Rowelett provided the satellite image that I was able to use to find Mibya Kyun. I sent him a picture of the lighthouse, and he posted it on his site: https://www.ibiblio.org/lighthouse/mmr.htm.

278    *women's issues are often split*   NiNi shared a copy of her speech: Prof. Dr. Khin Ni Ni Thein, "Women's Empowerment and Leadership for Sustainable Water Governance and Development: Myanmar Case on Behalf of Women for Water Partnership (WfWP)," 2014.

## A Living Link

281    *his daughter leaves behind a tree*   In the books on Ashoka referenced above, I was seeking more information on my namesake Sangamithra, spelled *Sanghamitta* or *Sanghamitra*. An artistic sketch of Sanghamitra with the Bodhi tree composed by Nandalal Bose is provided in Nayanjot Lahiri, *Ashoka in Ancient India* (Harvard University Press, 2015).

## Trying

An earlier version of this chapter was published here: Sangamithra Iyer, "Trying," *Newtown Literary* 12, Spring/Summer 2018.

## Sanctuary City

This section was supported by a residency at the Camargo Foundation funded by the Jerome Foundation. A short excerpt was published in French, available at https://www.journalventilo.fr/wp-content/uploads/filebase/1001nuits-saison2.pdf.

295    *The neighboring city of Marseille*   This is a great source on the history of industrial pollution in Marseille: Xavier Daumalin, *Les Calanques industrielles de Marseille et leurs pollutions—une histoire au présent* (REF2C, 2016).

295    *The Endangered Species Act defines* **take**   This reflection on the ESA was also included in my Lost

Species Day talk, "On the Brink," hosted by Extinction Rebellion and Writers Rebel on November 30, 2020.

296   *spews xenophobic vitriol*   Colin Dwyer, "'Racist' and 'Shameful': How Other Countries Are Responding to Trump's Slur," *The Two-Way, NPR*, January 12, 2018, https://www.npr.org/sections/thetwo-way/2018/01/12/577599691/racist-and-shameful-how-other-countries-are-responding-to-trumps-slur.

296   *and bans bodies*   Immigration and Ethnic History Society, "Muslim Travel Ban," The University of Texas at Austin Department of History, https://immigrationhistory.org/item/muslim-travel-ban/.

300   *a government spokesperson*   Norman Mailer, *The Armies of the Night* (Weidenfeld & Nicolson, 1968), 284.

301   *pilot on the TV series*   *30 Rock*, season 5, episode 14, "Double-Edged Sword," written by Tina Fey and Kay Cannon, directed by Ken Whittingham, aired February 10, 2011, on NBC.

302   *like Oliver said*   Mary Oliver, "Wild Geese," in *Dream Work* (The Atlantic Monthly Press, 1986).

303   **her assemblage of ordinary racial microaggressions**   Claudia Rankine, *Citizen: An American Lyric* (Graywolf Press, 2014).

306   *Sunu on a podcast*   Emily McGranachan, host, and Dakota Fine, cohost, *Outspoken Voices: A Podcast for LGBTQ+ Families*, season 3, episode 58, "Author, Poet, and Activist: Sunu Chandy," Family Equality, November 2, 2021, 40 min., 16 secs., https://familyequality.org/2021/11/02/58-author-poet-and-activist-sunu-chandy/.

305   *I review edits on my essay*   Sangamithra Iyer, "Are You Willing?," in *Writing for Animals: An Anthology for Writers and Instructors to Educate and Inspire*, ed. John Yunker (Ashland Creek Press, 2018).

305   *Here is a story / to break*   Mary Oliver, "Lead," in *New and Selected Poems: Volume Two* (Beacon Press, 2007).

A CATENA OF SOURCES

306     *a mortal fear of caring too much*   Naisargi Davé, "Witness: Humans, Animals, and the Politics of Becoming," *Cultural Anthropology* 29, no. 3 (August 2014): 433–56.

306     *I ask her to unpack this fear*   Naisargi Davé, "The Ethnography of Activism: The *Satya* Interview with Naisargi Davé," interview by Sangamithra Iyer, *Satya: The Long View*, 2016.

## Ungovernable Bodies

A few fragments from this section were excerpted as a broadside: Sangamithra Iyer, "Disenfranchised Grief in Four Parts," Thrash Press.

308     *three last northern white rhinos*   Gillie and Marc, *The Last Three*, 2018, bronze, 213 x 144 x 71 in. (540 x 365 x 180 cm), Astor Place, New York City, https://gillieandmarc.com/products/the-last-three-5.

311     *providing love and dignity*   Eileen Myles, *Afterglow: A Dog Memoir* (Atlantic Books, 2018). There is also this excellent conversation between Eileen Myles and David Naimon: David Naimon, host, *Between the Covers*, "Eileen Myles: Afterglow," Tin House, October 19, 2017, 1 hr., 28 min., https://tinhouse.com/podcast/eileen-myles-afterglow/.

311     *animal activists and goose advocates*   I want to acknowledge the work of David Karopkin and GooseWatch NYC.

311     *the sanctuary director tells us*   We attended a tour given by Rachel McCrystal, Executive Director of Woodstock Farm Animal Sanctuary.

312     *when you don't believe women*   More on Dr. Mona Hanna: https://monahannaattisha.com/.

318     *we discuss radiant friendships*   Sangamithra Iyer, Anu Radha Verma, and Anne Riley, "Radiant Practices: Relationship as Centre," moderated by Lisa Hirmer, ArtsEverywhere Festival, Guelph, Ontario, January 25, 2020.

318   ***her talk about trees***   More on Lisa Hirmer's writing on this subject: Lisa Hirmer, *Forests Not Yet Here* (Publication Studio Guelph, 2020).

323   ***If we say his name***   I wrote these lines attending from a writing prompt of a virtual reading, "Say Their Names: A Reading to Honor Black Lives," by Queens Writers Resist and Queensbound on June 24, 2020. Grateful for these communities holding space and creating space.

329   ***pilot whales that are stranded***   BBC News, "More than 100 Beached Whales Saved off Sri Lanka," *BBC News*, November 3, 2020, https://www.bbc.com/news/world-asia-54805138.

332   ***Extinction Rebellion literary event***   Extinction Rebellion and Writers Rebel, hosts, "On the Brink," virtual, November 30, 2020.

332   ***yellow booklet written about Jaya Aunty***   *Tulasi Mata*—see above, compiled by C. S. Rao.

## Speaker of Rivers

337   ***Reiko's tarot writing workshop***   I took Rahna Reiko Rizzuto's workshop as part of Two Trees Writers Collaboratory in October 2020.

## Collaboratory

339   ***a collaboration + laboratory***   I participated in The Collaboratory at the University of Guelph, Fall 2022, at the invitation of Dr. Madhur Anand at the Guelph Institute of Environmental Research.

## Bas Bas Bas Redux

341   ***Ganga filled with corpses***   Geeta Pandey, "Covid-19: India's Holiest River Is Swollen with Bodies," *BBC News*,

*Delhi,* May 18, 2021, https://www.bbc.com/news/world-asia-india-57154564.

341 ***American Veterinary Medical Association*** "AVMA Guidelines for the Depopulation of Animals," AMVA, 2019 edition, https://www.avma.org/sites/default/files/resources/AVMA-Guidelines-for-the-Depopulation-of-Animals.pdf.

342 ***They say Ventilator Shutdown*** Marina Bolotnikova, "Amid Bird Flu Outbreak, Meat Producers Seek 'Ventilation Shutdown' for Mass Chicken Killing," *The Intercept,* April 14, 2022.

342 ***Time to Silent*** Martin Sinel and Tony Weis, "Ventilation Shutdown and the Break-Taking Violence of Infectious Disease Emergency Management in Industrial Livestock Production," *Environment and Planning E: Nature and Space* 7, no. 3 (2024): 1076–97, https://doi.org/10.1177/25148486241229012.

## Shizengaku

343 ***collecting monkey specimens*** Jack Fooden, "Systemic Review of Southeast Asian Longtail Macaques," Chicago Field Museum of Natural History, 1936.

344 ***Three monkeys escape*** Elizabeth Gamillo, "All Animals Are Accounted for After Truck Carrying 100 Lab Monkeys Crashed in Pennsylvania," *Smithsonian Magazine,* January 26, 2022, https://www.smithsonianmag.com/smart-news/monkey-business-escaped-laboratory-monkeys-after-pennsylvania-crash-found-180979460/.

344 ***The translator notes*** Takayoshi Kano, *The Last Ape,* trans. Evelyn Ono Vineberg, (Stanford University Press, 1992), xiv.

345 ***monkey washing a sweet potato*** Tetsuro Matsuzawa and William C. McGrew, "Kinji Imanishi and 60 Years of Japanese Primatology," *Current Biology* 18, no. 14 (2008).

345   ***Imanishi writes this short book***   Kinji Imanishi, *A Japanese View of Nature: The World of Living Things by Kinji Imanishi*, trans. Pamela J. Asquith, Heita Kawakatsu, Hiroyuki Takasaki, and Shusuke Yagi, ed. Pamela J. Asquith (Routledge, 2002).

345   ***this new word in my mouth***   Imanishi, *A Japanese View of Nature*, xxvii–viii.

346   ***on similarities and differences***   Imanishi, *A Japanese View of Nature*, 2–3.

346   ***have a wholeness and autonomy***   Imanishi, *A Japanese View of Nature*, 32.

346   ***the concept of affinity***   Imanishi, *A Japanese View of Nature*, 4–5.

346   ***he recognizes other knowledges***   Imanishi, *A Japanese View of Nature*, 7.

## ViaViaVia

349   ***I write to my kalyanamitra***   For more on Prof. Dr. Khin Ni Ni Thein (NiNi), founder and chair of Myanmar Water Academy, and her work, visit https://www.myanmarwatersacademy.com/.

## Sangam of Sangams

352   ***These poetic infrastructures***   I am grateful to Lisa Hirmer for inviting me to participate in "Careful Infrastructures for These Warming Days."

352   ***I learn the Kattunayakans***   Helina Jolly, Terre Satterfield, Milind Kandlikar, and Suma TR, "Indigenous Insights on Human-Wildlife Coexistence in Southern India," *Conservation Biology* 36, no. 6 (2022): https://doi.org/10.1111/cobi.13981.

SANGAMITHRA IYER is an environmental planner, engineer, and writer. She is the recipient of a Whiting Creative Nonfiction Grant, a Café Royal Foundation Literature Grant, and the Diamonstein-Spielvogel Fellowship at the New York Public Library.

An Emerging Writer Fellow at Aspen Summer Words, a finalist for the Siskiyou Prize for New Environmental Literature, and a recipient of a Pushcart Prize, Iyer has also received support from the Jerome and Camargo Foundations. She served as an editor of *Satya* magazine and as an associate for the environmental public policy action tank Brighter Green. Iyer is the founder of the Literary Animal Project, a habitat for conversations about how we portray animal lives on the page. She has devoted her career to watershed protection, wildlife coexistence, nature-based stormwater solutions, and sustainable cities.

Iyer holds a bachelor's in civil engineering from Cooper Union, a master's in geotechnical engineering from UC Berkeley, and a master's in creative writing from Hunter College. She lives in New York with her husband, Wan, and rescued pit bull, Asta.

## milkweed
EDITIONS

Founded as a nonprofit organization in 1980, Milkweed Editions is an independent publisher. Our mission is to identify, nurture, and publish transformative literature, and build an engaged community around it.

We are based in Bde Óta Othúŋwe (Minneapolis) in Mní Sota Makhóčhe (Minnesota), the traditional homeland of the Dakhóta and Anishinaabe (Ojibwe) people and current home to many thousands of Dakhóta, Ojibwe, and other Indigenous people, including four federally recognized Dakhóta nations and seven federally recognized Ojibwe nations.

We believe all flourishing is mutual, and we envision a future in which all can thrive. Realizing such a vision requires reflection on historical legacies and engagement with current realities. We humbly encourage readers to do the same.

milkweed.org

Milkweed Editions, an independent nonprofit literary publisher, gratefully acknowledges sustaining support from our board of directors, the McKnight Foundation, the National Endowment for the Arts, and many generous contributions from foundations, corporations, and thousands of individuals—our readers. This activity is made possible by the voters of Minnesota through a Minnesota State Arts Board Operating Support grant, thanks to a legislative appropriation from the Arts and Cultural Heritage Fund.

Interior design by Mike Corrao
Typeset in Freight Text

Freight Text was created in 2005 by the
African American type designer Joshua Darden.
Its design is inspired by the warmth and pragmatism
of eighteen-century Dutch typefaces.